HRM, TECHNICAL WORKERS AND THE MULTINATIONAL CORPORATION

White-collar jobs were once valued because they offered security, interesting work, a steady salary and the prospect of promotion. However, the relentless search for economic efficiency has apparently led to a situation where those holding a professional or managerial position must endure a range of policies designed to monitor, measure and reward individual performance while bearing the collective risk of being 'downsized' by distant employers who view them as the chief source of 'corporate fat'.

This book is the first major in-depth study of the impact of contemporary management practices on a rapidly expanding set of white-collar occupations, namely technical workers. It investigates whether HRM policies, such as employee appraisals and performance-related pay, have changed technical employment to the point where it is no longer based on a 'service contract'. In doing so, it provides a detailed examination of the nature of managerial control over employees who, by virtue of their commitment, present their employers with problems that are often ignored by prescriptive models of HRM.

The empirical evidence features case studies of matched pairs of hi-tech firms in the Irish Republic. The author examines recent debates about the nature of employment and the role of multinational corporations within the so-called 'Celtic Tiger' Irish economy.

HRM, Technical Workers and the Multinational Corporation will be essential reading for advanced students and researchers in human resource management, the sociology of work, organization studies and economic geography.

Patrick McGovern is Lecturer in Sociology at the London School of Economics and Political Science. His research on the employment practices of multinational corporations and human resource management has appeared in *Work, Employment and Society, New Technology, Work and Employment* and *Human Resource Management Journal.* His current research interests are the sociology of employment and economic life.

ROUTLEDGE ADVANCES IN MANAGEMENT AND BUSINESS STUDIES

1. MIDDLE MANAGERS IN EUROPE
Yves Frédéric Livian and John G. Burgoyne

2. MARKETING APOCALYPSE
Eschatology, escapology and the illusion of the end
Edited by Stephen Brown, Jim Bell and David Carson

3. RELATIONSHIP MARKETING IN PROFESSIONAL SERVICES
A study of agency – client dynamics in the advertising sector
Aino Halinen

4. JOB DESIGN AND TECHNOLOGY
Taylorism vs. Anti-Taylorism
Hans D. Pruijt

5. REGULATION AND ORGANISATIONS
International perspectives
Edited by Glenn Morgan and Lars Engwall

6. INFORMATION TECHNOLOGY, ORGANISATIONS AND PEOPLE
Transformations in the UK retail financial services sector
Jeff Watkins

7. HRM, TECHNICAL WORKERS AND THE MULTINATIONAL CORPORATION
Patrick McGovern

8. THE INTERNATIONALIZATION OF SMEs
The interstratos project
Edited by Antii Haahti, Graham Hall and Rik Donckels

Books published under the joint imprint of LSE/Routledge are works of high academic merit approved by the Publications Committee of the London School of Economics and Political Science. These publications are drawn from the wide range of academic studies in the social sciences for which the LSE has an international reputation.

HRM, TECHNICAL WORKERS AND THE MULTINATIONAL CORPORATION

Patrick McGovern

London and New York

First published 1998
by Routledge
11 New Fetter Lane, London EC4P 4EE

Simultaneously published in the USA and Canada
by Routledge
29 West 35th Street, New York, NY 10001

Typeset in Sabon by
BC Typesetting, Bristol
Printed and bound in Great Britain by
TJ International, Padstow, Cornwall

British Library Cataloguing in Publication Data
A catalogue record for this book is available from the British Library

Library of Congress Cataloguing in Publication Data
McGovern, Patrick, 1966–
HRM, technical workers and the multinational corporation/Patrick McGovern
p. cm
Includes bibliographical references and index.
1. White collar workers–Ireland. 2. Industrial technicians–Ireland. 3. International business enterprises–Ireland–Employees. 4. International business enterprises–Ireland–Personnel management. 5. High technology industries–Ireland–Employees. 6. High technology industries–Ireland–Personnel management. 7. Compensation management–Ireland. I. Title.
HD8039.M4I735 1998
658.3'044–dc21 97-52235
CIP

ISBN 0–415–18487–8

CONTENTS

FIGURES AND TABLES

Figures

Tables

ACKNOWLEDGEMENTS

This research was made possible by a Funded Studentship at Nuffield College, Oxford. My sincere thanks to the Warden, Fellows and secretarial staff for their support. The Prendergast Fund at the University of Oxford also contributed towards the costs of the fieldwork. A special thanks to the managers and employees of the four case study organisations for facilitating the fieldwork. Thanks also to the employees of Kinerton (Irl) for undertaking a pilot test of the questionnaire survey; to my parents who provided transport (among other things); and to my sister Carmel who spent many tedious weekends helping to transcribe interviews.

I had the privilege of receiving much-needed advice and constructive criticism from Roderick Martin and Duncan Gallie throughout the course of the project. Others who commented on various parts of the research at various stages were: in Oxford, Jens Bastian, Tak Wing Chan, Jon Clark, Colin Crouch, John Geary, Brendan Halpin and May Tam; in Dublin, Teresa Brannick, Bill Roche and James Wickham; in Birmingham, Ossie Jones; and in London, Colin Mills. None of the above should be held responsible for any of my mistakes. Finally, I would like to dedicate this book to Katrina McLoughlin for her help, love and understanding.

Parts of Chapters 3, 5 and 7 are from previously published journal articles. I would like to thank the editors and publishers of *Human Resource Management Journal* (Industrial Relations Services and Personnel Publications Ltd), *Work, Employment and Society* (BSA Publications Ltd) and *New Technology, Work and Employment* (Blackwell Publishers Ltd) for permission to use this material.

1

MANAGEMENT PRACTICES, TECHNICAL WORKERS AND THE SERVICE RELATIONSHIP

Introduction

Will white-collar employment ever be the same again? Anybody reading the business pages during the recession-bound years of the early 1990s might conclude that it had changed beyond recognition. The emergence of corporate downsizing and delayering appeared to shatter white-collar assumptions about job security. The simultaneous introduction of various performance measures, such as individual appraisals and league tables, added pressure onto those who had begun to feel insecure. Managers were being made redundant on a large scale while those who survived found that their roles had enlarged because there was nobody else to take on the work of those who had departed. The solution appeared simple because there were no alternatives: cling on and hope for early retirement.

In this context there has been considerable speculation that the days of a career with a single employer are at an end. The best-selling author Charles Handy (1989) has long since promoted the view that managers are moving towards a portfolio career that consists of a small number of related jobs with different employers. The Harvard Business School academic, Rosabeth Moss Kanter (1989), has added further weight to this claim by arguing that the managerial career would in future be based on hopping between organizations rather than climbing up within any one specific organization. If this picture appears to be unduly exaggerated, it can be supported by evidence culled from an expanding body of social science research into, what Heckscher (1995) calls, the 'white-collar blues'.

Heckscher's (1995) study of eight major US companies concluded that the recent wave of corporate downsizing had amounted to a systematic assault on middle management. Corporations no longer offered lifetime security. Managers, for their part, abandoned a tradition of unquestioning loyalty since they now knew that this would not be reciprocated. Those who coped best in this environment were those who had given up any hope that the organization would settle down or return to normal. Other recent studies of white-collar employment in the US also concluded that the prospects for job security and upwardly mobile careers had been significantly reduced as a result of industrial restructuring (Osterman 1996). Similarly, surveys in the UK by the Institute of Management reported that 80 per cent of individual managers had experienced restructuring at least once in the previous five years (Wheatley 1992) and a similar percentage were worried about the lack of promotion opportunities (Lockwood *et al.* 1992). More significantly, a study of employers in the south of England by Brown and Scase (1994) reported that 'the demise of the bureaucratic paradigm' and the development of flatter 'adaptive' organizations had transformed the traditional career to the point where graduate recruits could no longer expect to climb into higher managerial positions. Finally, detailed case studies of the impact of downsizing on the managerial career in six major British-based employers reported that expectations of job security and of promotion had been reduced dramatically (McGovern *et al.* 1998).

The focus of this book, however, is on another set of changes. While they may have had a lower mass media profile they are no less significant for that. They may even be more widespread. These relate to a process which David Lockwood (1958) called the 'bureaucratization of employment' in his classic study of office clerks. This is defined as 'the rationalization of work relations which is achieved mainly through the establishment of universalistic criteria governing conditions of employment, and through the specification of tasks within a clear-cut hierarchy of authority' (Lockwood 1989: 231). The bureaucratization I am interested in stems from the increasing use of formalized management policies such as project management systems, performance appraisals, performance-related pay systems and other measures. While all of these are intended to help management to manage in large impersonal organizations, they have other effects that are of sociological interest. They make formal what was previously informal by, for instance, introducing systems of monitoring and evaluation where none existed previously. They also bring a much greater degree of formal rationality, predictability and reliability into the world of work. In other words, they extend the classical bureaucratic principles of rationalization described by Weber (1948). The spirit of these developments has been captured in Ritzer's (1993) influential repackaging of Weber's theory as one of 'McDonaldization'. This describes a process whereby the principles of the well-known fast-food restaurant come to dominate more and more areas of social life because of their ability to offer the producer increased efficiency, predictability, calculability and control. The world of employment is not exempt from these developments. Indeed it contains some of the most developed forms of this tendency. Ritzer's examples include Frederick Taylor's (1947) principles of scientific management, Henry Ford's assembly line and the Japanese 'Just-in-Time' production supply system.

While the spread of Weber's iron cage of rationality is of interest in itself I am more concerned with the impact that this kind of process has on employment relations within white-collar work. More specifically, I wish to assess whether contemporary management practices have transformed white-collar work to the point where it resembles that of blue-collar waged labour in a number of major respects. There is of course already a significant body of research that examines how bureaucratization has led to the degradation of work (Braverman 1974). But as Lockwood (1989) has argued, bureaucratization does not necessarily lead to unskilled, repetitive work with low pay and no job security. What it describes is a process by which work relations in general and employment conditions in particular are increasingly rationalized. In comparison with the vast amount of literature on the deskilling of white-collar work produced by those working within the Marxist labour process tradition (e.g., Crompton and Jones 1984; Armstrong *et al.* 1986), studies of the bureaucratization of white-collar work have been comparatively rare (Lockwood 1989: 231). This may be the result of an assumption that white-collar employment has already been bureaucratized and is therefore of little interest. The danger with this assumption is that bureaucratization is accepted as a *fait accompli*.

This study breaks new ground in addressing this issue in the context of human resource management practices. Furthermore, the answer to this question has important implications for existing conceptions of social class. Social class is of course one of the major areas of research in sociology. Much of the recent British debate has sought to understand the impact of social change on the middle classes (Savage *et al.* 1992; Butler and Savage 1995). Here also the debate has suffered from a lack of empirical evidence on changing employment practices. Despite the significance attached to new management practices in this debate, little attempt has been made to undertake detailed employer-based studies of these changes. This is surprising since an individual's employment status is one of the key indicators of social class location.

While this debate informs the research presented in this book I should like to emphasize that this is primarily a study in industrial sociology rather than in social class. The reason I draw so heavily on the sociological literature about the middle class is that it offers a powerful theoretical account of the employment relationship for professional and managerial workers. Furthermore, some of the questions I address are derived from issues raised within this literature. It would be wrong to imply from this that I do not intend to shed some light on the debate over the changing middle classes. However, the focus on managerial practices in professional white-collar work, and the overall focus, are such that this is a study in the sociology of work and occupations rather than of social stratification.

This may seem somewhat old-fashioned in an era when research in the sociology of employment has been crowded out by management-oriented studies that bear the title of human resource management (HRM). However, some of the more interesting work in this area has been conducted by scholars who write from a sociological perspective such as John Storey (1992), Peter Armstrong (1995) and Stephen Wood (Wood and Albanese 1995). Nevertheless, much of the substance – the theoretical models and the research programmes – is dominated by a management perspective. This seeks to establish models of 'best practice' and in extreme cases represents an ideological project that seeks to provide the ultimate control of employee performance through the management of meaning (Keenoy and Anthony 1992: 248). In this context, I believe there is an even greater need for a sociological perspective that examines how contemporary management practices shape the social construction of the employment relationship. This is all the more necessary because management-oriented approaches either do not examine these issues or else provide only a partial understanding.

In the next part of this chapter I shall review some of the existing theories and evidence on the changing nature of white-collar employment. My aim, in this review, is to show that these debates have suffered from a lack of detailed organizational level data, especially in regard to 'new' management practices. Particular emphasis will be placed on Alan Fox's (1974) influential micro-sociological model of the employment relationship which seeks to characterize it as one of varying degrees of trust. Fox's (1974) work is an appropriate starting point because it provides a coherent theoretical account of the white-collar employment relationship. The value of Fox's account has also been recognized by other sociologists who have incorporated it into recent studies of specific groups of white-collar workers such as engineers. More specifically, it is of interest because of arguments which claim that the traditional element of trust has been eroded from the employment relationship for professional and managerial workers. After describing this particular argument I shall focus on some contemporary management practices, especially those that lie within the area of HRM, since these are, supposedly, the source of this change.

Trust and the employment relationship

Fox's theory of trust and work roles

The model which Fox (1974) set out in *Beyond Contract* for the analysis of different patterns of work organization, was presented as a 'necessary foundation' to the subsequent examination of the non-contractual elements of the employment contract in capitalist societies. A central tenet of this model is the idea that variations in employment relations are related to variations in the discretionary content of work roles. Though Fox (1974: 16–21) acknowledged that work roles can be specific or diffuse in either the task-range sense, or the discretionary-content sense, his exposition is explicitly restricted to the latter. The task-range is, he claims, a lateral conception while the discretionary-content is more appropriately conceived of as a vertical conception. Fox adopts Jaques' (1967) distinction between 'prescribed' and 'discretionary' work as a means of analysing variations in the discretionary-content of work roles. While all roles contain both elements, the discretionary-content of jobs generally 'tends to enlarge further in scope and importance' as one moves upwards in the occupational hierarchy (Fox 1974: 24).

Three broad categories of work roles are described. Each of these syndromes of low, medium and high-discretion is strongly associated with low, medium and high levels of trust. In Fox's terms, trust is concerned 'with relationships which are structured and institutionalized in the form of roles and rules' (Fox 1974: 68–69), rather than with personal relations between individuals as in the conventional usage of the word. Each of these categories is also associated with a particular set of work relations. High-discretion high-trust work patterns are characterized by a high degree of moral involvement (Etzioni 1961) in the organization and also by a reciprocal willingness on the part of employers and employees to undertake favours for each other. Such relations, which are entitled 'social exchange', characterize the employment of managers, engineers, scientists and other professionals. Low-discretion low-trust work patterns, by contrast, are characterized by 'economic exchange'. Those who occupy low-trust positions are, in effect, being treated by management as if they cannot be trusted to act in the employer's interest of their own accord. Their exchanges with management are characterized by an instrumental rationality because of the widespread perception that management has no need to trust its employees because of the detailed division of labour. This contributes to a low-trust dynamic in which both managers and workers act in a cautious calculative manner.

It is important to emphasize that Fox's conception of trust is not concerned with personal relations between individuals, 'but with relationships which are structured and institutionalized in the form of roles and rules' (1974: 68–69). This gives Fox's use of the term a rather idiosyncratic quality: in normal usage trust refers to a belief or expectation about the dependability or honesty of another person. Studies that have attempted to measure the amount of trust which workers have in management have recognized this limitation and have focused on the other components mentioned by Fox such as the levels of discretion and supervision in work (e.g., Whelan 1982: 31–38).

Even so, the level of discretion in work may not be interpreted by workers as a direct reflection of trust in them by management. This argument has been advanced by Roche who, in a critical evaluation of Fox's theory, argues that trust perceptions may be more usefully regarded as a feature of *social integration* rather than *system integration*. Perceptions of trust, he believes, may be of particular consequence at an interpersonal level, e.g., between individual negotiators, and especially in situations of organizational and procedural disruption (Roche 1991: 109). They are of little value in explaining relations between institutions at the macro level.

Nevertheless, Fox's analytical categories provide a number of hypotheses for empirical research even if his claims about the links between job content and worker perceptions are flawed (Roche 1991). Of particular interest here is the claim that employment relations in the higher echelons are characterized by open-ended exchanges with a high degree of employee commitment. This claim, which describes those employed in the category of high-trust high-discretion work roles, was itself based on a wide-ranging synthesis of the existing research on organizational professionals (1974: 30–37). But does this conceptualization of professional and white-collar work still stand?

Trust and 'service-class' employment

While Fox's discussion of the employment contract and employment relations continues to be of significance (e.g., Streeck 1992: 41–75), his original micro-sociological theory of trust and organizational integration received limited attention in the decade following its publication (apart from Wood and Elliot 1977). This situation has changed in more recent times as Fox's conception of high-discretion high-trust roles has been utilized in the analysis of work organization (Heisig and Littek 1995), explanations of the organizational integration of white-collar employees (Hyman 1989: 34) of managerial employees (Armstrong 1989) and, in particular, in studies of technical workers (e.g., Whalley 1986a; Armstrong 1987; Causer and Jones 1990; Webb 1992). However, it is perhaps through Goldthorpe's (1982) influential essay on the formation of the *service class* and the subsequent debate over its changing nature that Fox's arguments have become most widely known (see also Lash and Urry 1987).

The service class, as described by Goldthorpe (1982), consists of higher white-collar employees (though not the self-employed or employers). Their employment is characterized by relations of trust or 'social exchange' of the kind described by Fox. According to Goldthorpe those in positions of authority, or positions where specialist knowledge is required, are given 'some legitimate area of autonomy and discretion'. Their work is therefore marked by a distinctive degree of autonomy especially in relation to the way that their performance is monitored. In these cases 'it must then *pro tanto* be a matter of trust that they will act – i.e., will make decisions, choices, judgements, etc. – in ways that are consistent with organizational goals and values' (1982: 168). It follows that the performance of these employees will be driven by their moral commitment in the organization, rather than by external sanctions and rewards (see also Goldthorpe *et al.* 1968: 39–40).

Relations of trust are, however, only one element of Goldthorpe's account of service-class employment. Another defining feature, according to Goldthorpe is the way that the market character of this service relationship is modified in a qualitatively different way to that of the wage-worker. Exchanges between employer and professional or managerial employees are more diffuse, have a more long-term orientation and a higher moral content. Payment is made as a form of 'compensation' and 'consideration' for discharging trust faithfully. This is typically in the form of a salary with some fringe benefits. Here Goldthorpe emphasizes the role 'played by rewards that are of an essentially *prospective* kind: that is, embodied in understandings on salary increments, on security both in employment and after retirement and, above all, on career opportunities' (1982: 169). These characteristics act to distinguish members of the service class from wage-earners who typically have a labour contract that is based on the exchange of discrete quantities of effort for discrete quantities of reward on a short-term basis. In sum, the defining element of professional and managerial employment is one of 'service' while that of the wage-worker is based on the logic of contractual exchanges (see also Goldthorpe 1980: 39–42; Erikson and Goldthorpe 1993: 41–42). This argument has become a fundamental feature of Goldthorpe's influential programme of research on social mobility (Goldthorpe 1980; Erikson and Goldthorpe 1993).

In his initial formulation of the service class, Goldthorpe drew explicitly on Lockwood's (1958) use of the Weberian concepts of an employees' market and work situations since these were, in his view, the two major components of class position. 'Market situation' is defined by Lockwood (1958: 15) as 'the source and size of income, degree of job security, and opportunity for upward occupational mobility'. 'Work situation' refers to 'the set of social relationships in which the individual is involved at work by virtue of his position in the division of labour'. Goldthorpe argued that it was possible to combine these concepts because those in the *same occupation* typically shared similar market and work situations (1980: 39). This meant that the amount of autonomy in an employee's work role was a direct indicator of his or her location in the class structure. In the case of the service class Goldthorpe argues that 'the conditions – in effect of, bureaucratic employment – are ones which clearly reflect, whether by design or evolution, the need for creating and sustaining an organizational commitment' (1982: 168). In other words:

> A service relationship can thus be understood as the means through which an employing organization seeks to create

> and sustain such [moral] commitment; or that is, as a functional alternative to direct control in regard to those employees whom the organization must to some significant extent *trust* to make decisions and to carry them through in ways that are consistent with organizational values and goals.
>
> (Erikson and Goldthorpe 1993: 42)

Goldthorpe (1995) has since revised his earlier model of service-class employment in order to respond to criticisms made by Marxist and other writers. He still believes that those who hold positions governed by a service relationship will tend to be those who also experience a considerable degree of autonomy in their work roles. However, he now claims that where there is a discrepancy between these two then what really matters is the nature of the employment relationship and not the amount of autonomy or discretion as Marxist writers claim (e.g., Crompton 1980). He reiterates the point that the key question is whether or not a relationship of service exists. Such a relationship is typically demonstrated by employment that contains prospective elements – salary increments on an established scale, an understanding that job security can be expected along with the real possibility of an organizational career. This revision is in line with the earlier interpretation of his position by Marshall *et al.* (1988: 13) to the effect that it is the *conditions* of employment and not the functions of employment that determine class location.

While there is much that is of analytical value in Goldthorpe's arguments, his decision to discard the work situation, because of the possibility of a discrepancy between conditions of employment and work tasks, seems premature if convenient. It is difficult to see why such a revision is necessary when there is no evidence of a widespread discrepancy. If anything, the evidence presented by Evans and Mills (1997; 1998) points in the other direction. Using data from large-scale national surveys in Britain they have examined whether the distinctions made by the Goldthorpe class schema actually fit the job characteristics of various types of employees. They found that the job characteristics could be divided into three major groups which corresponded with Goldthorpe's distinctions between service, intermediate and waged labour contracts.

Even if there is no divergence between work and market situations it is still necessary to explain how this is reflected in the work situation. More specifically, we would need to know how the various types of contracts are reflected in the way employers structure jobs and occupations and in the way they administer their authority. These are issues that have a long and fruitful tradition in the sociology of work. Industrial sociologists might therefore argue that we do not need to draw on debates over social class in order to provide a rationale for tackling such issues. While I would agree with this in principle, a number of recent studies of the professional and managerial middle classes have presented arguments that have implications for our understanding of white-collar employment *per se*. These claim that significant changes have taken place in the way that these employees are managed.

The erosion of trust in service-class employment?

What is of most concern here is the argument that traditional elements of service-class employment have been eroded by the introduction of new management practices. This argument has appeared in studies of the service class, in general, and of technical workers, in particular. Both will be reviewed here in detail since they are of direct relevance to this study.

The fragmentation of the service class

Savage *et al.* (1988; 1992) argue, on the basis of research conducted along the so-called 'M4 corridor' in the South-East of Britain, that the service class is becoming increasingly fragmented. In contrast to Goldthorpe's claim of increasing cohesiveness, they find that a cleavage has emerged between those professional and managerial workers based in bureaucratic employment, and the more entrepreneurial professional groups who either move between employers or are self-employed. In particular, they argue that those in managerial positions have not fared as well as professionals with regard to social mobility. This is because their labour market chances depend on, what Wright (1985) has termed, 'organizational assets'. According to Savage *et al.*, the problem with organizational assets, such as authority, is that they do not have the same market capacity as the academic or vocational qualifications that professionals possess. Consequently, managers find it more difficult than professionals to advance their careers by 'job hopping' across organizations.

For our purposes, the most interesting feature of their research is their evidence on, and interpretation of, the role of management policies. With regard to those in bureaucratic employment, they claim that employers no longer have an open-ended trust in these employees:

> (the) type of salary is changing with the growth of complex systems of appraisal and merit payment, so that salary becomes more closely linked to the direct productivity of the professional or managerial worker. Employers do not simply 'trust' that these workers will get on with it: they devise measures to ensure their actual work performance can be evaluated and their rewards calculated accordingly.
> (1988: 463)

It is clear from this extract that they attach considerable significance to the finding that the payment of organization-based professional and managerial employees has become increasingly related to their performance and skills, rather than simply to their position in the organizational hierarchy. In other words, the service contract has changed since the principle behind the method of remuneration has changed. This is no longer solely in the form of compensation or consideration for the faithful discharge of responsibilities. It has instead become much more closely related to individual performance. Other indicators of this general change include 'clocking in' arrangements for top-level managers, single-status canteens and open-plan offices.

It is worth pointing out here that the evidence for these changes is not convincing. They appear to be drawn from interviews with four companies who had their Head Office in the Slough area (Savage *et al.* 1988: 460). It is not clear exactly who, or how many people were interviewed, or indeed what precisely they were asked. Such details are important in enabling us to accept the significance that the authors attach to various management practices. In the case of appraisals and performance-related pay, it is not clear that these practices are present in *any* of the four firms! It is, accordingly, impossible to know what proportion of pay is based on performance, what impact this has on employees or how successfully these practices are implemented. Furthermore, the introduction of 'clocking in' arrangements for all employees – including managers – and the use of a single-status canteen seems to be solely the work of 'Firm W' (1988: 463). While a larger number of firms would appear to have introduced open-plan offices it is difficult to see how this has marked a decline in the autonomy of service-class workers. Presumably, autonomy is characterized by having something to hide!

This initial argument about management practice was supplemented by a more general claim when the authors discussed these issues in their book *Property, Bureaucracy and Culture* (Savage *et al.* 1992). In this they argue that bureaucratic hierarchies are becoming less significant in economic organizations. They state that: 'Organizations decreasingly use managerial hierarchies to structure their activities, but attempt to introduce forms of market mechanism instead' (1992: 212). Consequently the career strategies of organizational professionals and managers, and the type of work that they undertake, appear to be departing from those associated with more traditional bureaucratic hierarchies. They conclude that there is now an increasing trend towards inter-organizational mobility among service-class workers. This in turn is driven by a decline in opportunities for internal promotion, especially for middle managers (see also McGovern *et al.* 1998). However, it should be noted that much of the evidence for this decline is based on case studies and secondary data. The full extent of these changes has yet to be revealed.

Contemporary management practices and technical workers

The nature and significance of changing managerial practices have also been addressed in the growing literature on professional technical workers. Of the more significant contributions to this literature, the most ambitious is the trilogy of workplace studies of engineers by Zussman (1985), Whalley (1986a) and Crawford (1989). Each sought to test competing theories of professionalization and proletarianization using comparisons between engineers employed in old and new industries in Britain (Whalley), America (Zussman) and France (Crawford). Each also drew on Fox's (1974) analytical model of trust relations and Goldthorpe's approach to the service class.

Of these, Whalley's (1986a) study of two British factories, Computergraph and MetalCo, is the most explicit in its depiction of the employment of engineers as one characterized by high levels of trust (see also Whalley and Crawford 1984). He describes British engineers as 'trusted employees, part of a wider service class in industrial capitalism' (1986a: 13). The integration of engineers into these firms is achieved by the high levels of trust bestowed on the engineers by management. This is indicated by the level of autonomy contained in their work: they exercise considerable discretion when designing valuable machinery, preparing estimates and negotiating with manual workers about their daily practices. Though it might vary between R&D and the drawing office this, according to Whalley, still separates engineers from the shopfloor operatives. This argument leads to the conclusion that 'Engineers, . . . are not trusted by management because they are engineers, they are engineers because they are trusted' (1986a: 70). This phenomenological approach claims that occupations are *constructed* within the workplace. In this case, engineers are constructed through the amount of trust placed in them by management. This position can, however, be criticized for failing to account for the way the educational system constructs engineers outside of the workplace. It does so by awarding qualifications that signify their specialist knowledge of electronics etc. Contrary to what Whalley's case studies indicate, Britain is not unusual in this practice (Jones *et al.* 1993). Engineers can also be differentiated from shopfloor workers by the higher status associated with their abstract knowledge and, consequently, by the relatively higher levels of organizational responsibility bestowed on them by management (McGovern 1996a).

Nevertheless, Whalley's research does not point to an erosion of trust by management. Rather his argument supports the idea that the work situation is one of open-ended trust though there may be variations in the way these trusted workers are integrated into the organization (Whalley 1986b). For instance, the engineers in Computergraph were insulated from market pressures and could enjoy their work without having to worry unduly about the financial consequences. In MetalCo, by contrast, the engineers were directly exposed to product market pressures as their activities were organized around profit centres.

Whalley's approach has been criticized by Smith (1990) who argues, from a Marxist perspective, that the concept of *trusted workers* is of limited analytical value. Whalley's argument, according to Smith, over-plays the stability of the autonomy and trust associated with engineers because it ignores the contested nature of the exchanges between capital and labour. Accordingly, Smith (1990: 246) claims that Whalley misses the important point that 'technical workers enter into conflict with capital, regardless of cultural or organizational design, because of pressures by management to increase their productivity or exploitation'. From this he concludes that these workers should be viewed as 'waged labour' and not as 'a *permanently autonomous* and trusted group'.

Smith's (1990) argument would be more persuasive if he had described the kind of management practices he has in mind. While he has undertaken detailed case study research on technical workers in two firms (British Aerospace, Filton and Cadbury's, Birmingham) his argument tends to rest on a general differentiation between internal controls (career structures, staff status) and control through exposure to product market forces. To be fair, his research was not strictly concerned with management practices in technical work. However, his earlier study of British Aerospace concludes that these workers are clearly engaged in a labour process that is subject to various measures of effort intensification: 'it is capital which exploits technical workers, rationalizes and intensifies their work and generally shapes their experience of wage labour' (1987: 299). Again, there is little evidence as to how this occurs in practice. In any case, as Edwards (1990: 11) argues, Smith does not consider how far these technical workers were a special case of 'waged labour' since they worked in a factory with a strong craft union tradition. Nor does he acknowledge the limitations of basing his argument on a single organization. Also, he only spoke to members of TASS, a trade union with a tradition of militancy that has many similarities with the practices of skilled manual workers' unions.

Whittington (1990; 1991a; 1991b) has also claimed on the basis of his case study research on fourteen British R&D organizations that the traditional levels of autonomy and job security enjoyed by scientists and engineers has been eroded. This, he believes, is part of a wider pattern in which large corporations respond to increased market pressures by imposing market disciplines on their labour (1990: 200). It can also be attributed to a process which is leading to the fragmentation of the 'service class' as a whole (1991a: 103). Prior to this dramatic change he claims that professional scientists and engineers worked in an environment that was relatively free of market controls. This approach persisted for many years by virtue of the creative nature of R&D work and through the professions' distinctive ideological claim to autonomy. With the 'externalization' of R&D work, Whittington claims that this autonomy has been reduced by the pressures of increasing global competition and the growing threat from the market for corporate control.

However, Whittington's research is rather vague about how the change is reflected in managerial practice. This is all the more surprising when one considers that all of his forty-one interviews were conducted with managers (1991a: 94). What appears to have changed is the context and with that a new emphasis on accountability to the market. A more detailed account of contemporary management practices in technical work is provided by Causer and Jones (1990; 1993; 1996). Drawing on in-depth research into the employment practices of twenty electronics firms located in central southern England, they claim that hi-tech employers have four principal mechanisms for regulating the work of engineers:

(1) the organization of technical work;
(2) the supervision of technical work;
(3) career structures; and
(4) the pay and appraisal process.

In relation to the first, Causer and Jones (1990) found that virtually all research, development and design work was carried out in teams. The composition of these teams varies from project to project and individual members of any project team are rarely working on that project alone. It is worth adding here that Crawford's (1989: 89) study of French engineers also stresses the project nature of much engineering work, a characteristic which he suggests is typical not only of particular functions such as research and development, but of high technology industries in general.

Causer and Jones placed considerable emphasis on supervision as a means of managerial control. In the smaller departments in electronics firms, the supervisory process is characterized by 'fraternalist' strategies where managers work alongside their subordinates 'on a substantially egalitarian basis'. In larger departments this process is characterized by 'collegialism' where the manager/supervisor has a limited technical input on a day-to-day basis yet likes to offer advice based on his or her background. Discussions of technical problems are conducted with 'colleagues' rather than with 'subordinates', that is, on the basis of technical expertise rather than on notions of power. Both styles of supervision allowed the engineers a certain amount of autonomy while ensuring that projects were carried out on time and to a high standard.

The payment systems used by these employers allow for a substantial discretionary component. Though the performance appraisal system is normally separated from the pay review it is the key measure of the contribution of individual members of staff (see also Randle and Rainnie 1997). The problem of evaluating individual contributions is overcome by having section or project leaders conduct the appraisals. These typically have sufficient technical experience and knowledge of the work to carry out appraisals. Many of the firms provided a dual career ladder that met the career needs of those who were only interested in a technical career as well as those who wished to move into management.

More recently, Causer and Jones (1996) have highlighted two mechanisms that they consider to be of particular importance in managing these employees. They believe that a measure of control is secured through the development of organizational positions that combines elements of both technical and managerial work and through the operation of personalized payment systems.

Causer and Jones argue that each of these mechanisms serves to retain and to motivate technical staff. Of particular interest to this discussion is their conclusion that the engineers covered in their study were in many respects the archetypal 'trusted workers'. They conclude that the work of these employees is indeed characterized by a significant degree of autonomy and discretion. However, they add that this 'does not indicate an absence of managerial control' but rather the existence of 'certain types of control mechanisms appropriate to this kind of technical labour' (Causer and Jones 1990: 26).

Despite their limitations the research by Savage *et al.* and Causer and Jones indicates that the high-trust high-discretion conceptualization of organizational professionals does not fit in the context of contemporary human resource management practices. While other areas of general management practice, such as the use of project teams, also feature, it is those that relate to the reward element (pay and careers) of the effort-reward bargain (Baldamus 1961) that are of most significance. Yet the idea that HRM practices should reduce rather than enhance trust runs counter to some of the literature in this area.

HRM and high-trust work relations

The emergence of HRM, and the subsequent debates over its nature, have been detailed elsewhere (Storey 1989; 1992; Blyton and Turnbull 1992; Legge 1995). Rather than review such debates, I wish to extract some brief points from this voluminous body to the effect that HRM contributes to high-trust work relations.

HRM, as distinct from personnel management or industrial relations, is presented as being the leading edge of labour management theory and practice (see Legge 1995: 62–95). If the thrust of some of the models of HRM (e.g., Devanna *et al.* 1984; Beer *et al.* 1985) is accepted then some of the more 'progressive' employers may seek to enhance the sense of service and duty associated with high-discretion roles. This should be true of what Storey (1992) and Legge (1995: 66–67) term the 'soft' version of HRM. Here employers seek to treat employees as valued assets who can be a source of competitive advantage through their commitment, trust, adaptability and high-quality skills and knowledge. Richard Walton (1985), for instance, stated in a much-cited article published in the *Harvard Business Review* that employers should change their employment practices from one of employee control to one of commitment. They should seek to empower these resourceful humans, to give them greater responsibility and participation in decisions relating to their work, especially as it is they who know what they are doing best of all. Furthermore, they should do so because it would improve the competitiveness of the business. This concern with employee participation and the generation of employee commitment is broadly equivalent to the idea of social exchange associated with Fox's (1974) conception of high-trust high-discretion work. This resemblance, especially the concern with the manner in which employees commit themselves to or are normatively integrated into organizations, has not gone unnoticed (e.g., Roche 1991: 100). A more explicit link has been through the use of Fox's term 'high trust' by Guest (1987; 1991) to distinguish between the stereotypical models of HRM and of personnel management. Employee relations under HRM would be based on unitarism rather than pluralism and on individualism rather than collectivism. They could also, according to Guest, be characterized as being 'high' rather than 'low' trust in nature.

These theories have not, however, been matched by management practice. The evidence from the Workplace Industrial Relations surveys indicates that there has not been any significant increase in 'progressive' management practices in British industry (Millward 1994). Although there is considerable case study evidence and some survey evidence of the adoption of *practices* associated with HRM in the UK, questions remain about the extent and purity of such practices (Legge 1995: 329). Storey's (1992) case studies of this chimerical phenomenon in fifteen different organizations found little evidence that the 'soft version' existed in a pure form. He did find that some HRM policies were 'bolted on' to the embedded system of industrial relations in some instances. In these cases, the blue-collar workforce were still managed according to the traditions of collectivized industrial relations while HRM policies on appraisal and reward were directed towards the white-collar staff. Another study by Truss *et al.* (1997) of eight major UK-based organizations also found no pure examples of the so-called 'soft version' of HRM. Instead they found that the rhetoric adopted by these employers frequently embraced the tenets of the soft model but their overall approach was marked by strategic control over the business by senior management combined with close attention to financial performance. The result was that business considerations always took precedence over human resource issues, notably in the area of training and career development.

Large-scale survey evidence from a national sample of UK manufacturing plants has also produced evidence of high-commitment practices. However, there was little evidence of a complete high-commitment model along the lines purportedly pioneered by IBM and Hewlett-Packard (Wood and Albanese 1995). Finally, a case study of Hewlett-Packard, one of the frequently cited examples of HRM 'best practice' found that even in this organization a detailed form of bureaucratic control (Edwards 1979) existed alongside the rhetoric of a high-commitment corporate culture (McGovern and Hope-Hailey 1997).

What these studies suggest is that the possibility of HRM enhancing the high-trust high-discretion elements of the employment relationship is most unlikely. The practice of HRM, as distinct from the prescriptive theory, is far removed from Walton's (1985) commitment strategy. Slight exceptions to this would appear to be foreign-owned companies (Lowe and Oliver 1991; Storey 1992; Wood and Albanese 1995). Otherwise British employers appear to be a mixture of what Rose (1994) termed 'improvisers and empiricists'. That is to say that there is little long-term employment planning and new practices are only introduced when it is clear that they have worked elsewhere. In sum, the existing research on HRM indicates that there is no general shift from control to commitment. From the case study evidence it seems that, at best, some new practices are 'bolted on' to old arrangements. Some of these may reduce the characteristic element of trust in the white-collar work situation. However, the effect may be exaggerated. This remains to be seen.

The focus of the study

Much of the argument about changes in professional and managerial employment conditions can be described, to some extent at least, in terms of a shift from hierarchy to market. Savage *et al.* (1992: 212) have, as indicated earlier, been explicit about this shift. The point I would like to make is that this shift to market criteria appears to have extended the process of bureaucratization. In Lockwood's (1989) terms, it has led to the further rationalization of work relations. For instance, employers are making increasing use of performance-related pay (Savage *et al.* 1988; Causer and Jones 1996); they are less concerned with providing job hierarchies to retain staff (Causer and Jones 1993); and service-class employees are increasingly 'job hopping' across the external labour market to advance their careers (Savage *et al.* 1988; 1992). The corollary is that employers are no longer relying solely on rewards of a prospective kind: salaries and the prospect of an upwardly mobile career with incremental increases in pay.

Presumably, these changes have arisen because employers find them to be more effective in extracting work from their employees. If this is the case, then it could be assumed that the constant search by capitalist enterprises for more effective means of regulating work will result in a dynamic which can only further erode the traditional elements of service-class employment. In this context, the practice of HRM can be interpreted as a new phase in the bureaucratization of employment. This assumes, of course, that employers can only progress by changing those employment practices which modify the market character of the service relationship in a way that makes it more similar to that of the wage-worker. Changes in this direction would see employers try to formalize arrangments whereby pay is exchanged for discrete amounts of work. Such moves would indicate that employers are moving from hierarchical to market criteria in order to regulate professional and managerial employment. This also assumes that employers are doing so because this is a more efficient way of regulating white-collar employment. But to what extent is this really the case?

I have argued that despite the importance of employment practices in this debate the evidence produced to address these issues has been of a poor quality. There has, for instance, been no attempt to examine employees' views on the significance of these developments. Also, the mere presence of certain practices, such as performance-related pay, is taken as evidence of a significant change in service-class employment (e.g., Savage *et al.* 1988; Halford and Savage 1995). There are at least three problems with this interpretation. First, it is not safe to assume that the existence of such policies implies complete implementation within organizations. On the contrary, there is evidence from case studies of eight major 'blue chip' British organizations which indicates that there is considerable variation in the quality and consistency with which managers implement some human resource management practices (McGovern *et al.* 1997). Second, in the case of performance-related pay, for example, little consideration has been given to its role within the remuneration package and within the employers' overall approach to pay. The evidence to date, which has placed considerable importance on this practice (e.g., Savage *et al.* 1988; Causer and Jones 1996), does not tackle such questions as whether or not there has been a distinct move to discrete amounts of performance for discrete amounts of pay or, more fundamentally, what the employers' intentions are.

This leads to a third problem in that it is not really clear that the presence of such policies actually alters what Goldthorpe (1995) considers to be critical: the 'service' nature of this form of employment. Here at least, Causer and Jones (1990) state that while there have been changes, electronics engineers are clearly in service-class employment. It should be noted, however, that they make this claim through a 'work-centred' argument that points to the levels of trust and discretion in the work of the engineers. Breen (1997) argues that the very nature of the service relationship is such that it is unlikely to be altered by such developments. This is because employers find it difficult or impossible to monitor the work of their employees since so much of it is intangible. Therefore they must rely on maintaining a substantial element of trust and discretion in the service relationship. However, Breen's argument, like much of the literature on service-class employment, is theoretically assumed rather than empirically demonstrated.

In sum, many of the questions raised by the debate over service-class employment require answers that can only be obtained through the kind of research which falls within the traditions of industrial sociology. What is needed is detailed research on the employment relations of service-class employees in the context of contemporary human resource management practices. It is here that this study seeks to contribute. Apart from the research undertaken by Causer and Jones (1990) there is, to my knowledge, no other study that has tackled these issues at the organizational level.

Research questions

This study examines whether, and to what extent, the employment of organizational professionals (and semi-professionals) has been subjected to the kinds of changes described earlier. This means that the dependent variable is the 'service relationship'. To address this question, the research will go beyond managerial rhetoric about the presence and operation of employment practices to see what impact they actually have on the 'service' relationship of technical workers. It will focus in detail on the policies and practices of employers and the experiences and attitudes of managers and their technical staff.

Technical workers were chosen for a number of reasons. First, technical workers are a relatively neglected group in the sociological literature (Smith 1987; Meiksins and Smith 1996). This is all the more surprising in the context of a shift to a post-industrial or service economy where technical workers play an increasingly important role within modern organizations (Barley and Orr 1997). Second, much of the literature that I address only relates to engineers and to electronics-based industries. I wish to extend this to chemists and technicians while also including chemistry-based industries. Second, technical professionals typically have a high level of commitment to their work (Kerr *et al.* 1977) and to their organization (Steers 1977; Mowday *et al.* 1979) though there may occasionally be a conflict between the two (Kornhauser 1962; Marcson 1969; Cotgrove and Box 1970; Raelin 1985). This is important because it indicates that these employees conform with the idea that service-class employment is characterized by a high degree of moral commitment.

Third, and most important, this research will be in a position to assess whether the changes described by Savage *et al.* for managers apply in the same way to professionals. According to their argument it is the organizational context that is of most importance. However, what differentiates professionals from managers is their possession of academic qualifications. This provides them with a highly transferable form of knowledge which, unlike that of managers, is much less firm specific. In other words, while the work situation of professionals and managers may be changing in the same direction, the market situation remains the same for those with qualifications. That is to say, there is a divergence between the work and market situations of organizational professionals. This assumes that employers place a particular value on qualifications. But what if employers wish to alter the hierarchy of qualifications? How does this affect the market situation of technical workers? Again, these are questions that require research at the organizational level.

Structure of the book

It follows from this that one of the key questions in this study concerns the nature of job hierarchies within firms. Is the notion of an organizational career breaking down as employers shift away from internal hierarchies towards market solutions? What is the employers' overall approach towards the recruitment and retention of technical workers? This subject will be examined in detail in two chapters. Chapter 3 investigates employers' recruitment practices and their role in trends towards qualification inflation in the labour market. Chapter 7 examines career structures and the employers' approach to the issue of employee retention. Much of the chapter will be taken up with the question of whether or not employers still build job hierarchies in order to reward and retain staff as well as to differentiate between their roles, experience and status.

The second major question concerns the organization of work and the nature of managerial authority in technical work. Chapters 4 and 5 examine the means of management control in technical work. Chapter 4 focuses on the role of formal bureaucratic and informal management methods for organizing and coordinating work. The second half of this chapter examines the fit between highly educated technical workers and the work offered by multinational corporations (MNC) branch plants. Chapter 5 describes the nature and role of human resource management policies and management style. Both of these chapters will explore Fox's (1974) argument that the nature of managerial control will be strongly influenced by the nature of the work and by employees' work orientations.

The changing nature of remuneration is examined in Chapter 6. There are two issues here. The first is whether, and to what extent, employers are moving towards performance-related pay. The second is the impact of these changes, if any, on the nature of employment relations. The principal findings of this book and their implications for the literature on social class and human resource management are discussed in Chapter 8.

2

MULTINATIONAL CORPORATIONS, HUMAN RESOURCE MANAGEMENT AND TECHNICAL LABOUR

Introduction

To address the questions identified in the previous chapter it was essential that the empirical research should be based on organizations which employed a significant number of technical workers and implemented a range of HRM policies such as appraisal and performance-related pay. The only employers that met both of these criteria in the Irish Republic were the branch plants of multinational corporations (MNC). The decision to carry out the research in MNC branch plants adds further complexity to the study as their role in newly industrializing economies in general, and their employment practices in particular, have been subject to long-running debates in academic and policy circles. But before taking up this issue I would like to make some brief points about the significance of the Irish context.

For those outside Ireland there may appear to be little to be learned from developments within this small European nation. This belief may possibly be reinforced by the paucity of research on the Irish experience of MNCs, especially that published outside of Ireland. However, there are at least three reasons why this particular study is of general import. First, the practices of MNCs should be of interest precisely because of the unusual hi-tech labour market in the Irish Republic. Contrary to the experiences of hi-tech employers in other countries (e.g., Turbin and Rosse 1990; Winstanley 1991), the labour market for technical staff in Ireland is characterized by skill surplus. This unusual situation provides a unique opportunity to assess much of the received wisdom drawn from studies of recruitment and employee retention in the context of skill shortages. Second, there is considerable evidence to show that US-owned MNCs tend to introduce employment policies from their home environment into their overseas operations (Kujawa 1979; Buckley and Enderwick 1985) with the result that there are considerable similarities between plants in different countries. To the extent that this is true, a study that focuses on these policies and practices is not bound by national context. Finally it should also be clear from the previous chapter that this research is concerned with sociological questions that cannot be explained by history or geography. It should not be inferred from this that the national context is irrelevant. Rather I believe that it is essential to examine employment relations in the context of different labour market conditions since this provides a basis for extending and refining theoretical arguments and models. In this way the explanatory power of various theories of employment relations can be tested and, where necessary, modified or rejected. In this instance the theoretical questions are taken from a debate about changes in British employment and applied within an Irish context.

This chapter commences with an overview of the arrival of multinationals in the Irish Republic. Their presence has been subject to considerable academic debate, especially in relation to Ireland's position in the new international division of labour. Some of the key issues raised by this debate will be outlined before describing the existing research on the human resource management practices of MNCs in Ireland. The remainder of the chapter will describe the case study firms, the technical workers and the methods of data collection.

Multinational corporations in the Irish Republic

According to O'Malley (1986), Ireland has experienced three distinct phases of foreign direct investment. The first phase (1930–1960) coincided with an era in which industrial policies were dominated by a philosophy of economic protectionism. The MNCs that arrived during this period were primarily British companies who had to establish manufacturing operations in order to gain access to the local market. Following a comprehensive review of economic policy, the Fianna Fail government of the late 1950s turned to foreign enterprise as 'pump-primers for an *Irish* dynamism' (Jacobsen 1994) that would haul the economy out of a prolonged period of stagnancy. During this second wave (late 1950s to 1970) much of the foreign direct investment was in labour-intensive operations such as clothing, footwear, textiles and light engineering industries. These enterprises, which included the first American MNCs, were attracted by the low labour costs and the prospects of a gradual shift away from the protectionist economic measures of earlier decades (Kennedy *et al.* 1988).

The third phase, which began in the early 1970s and continues to the present day, is marked by even greater reliance on MNCs principally because indigenous industrialists failed to show the required levels of dynamism. In this period the state's economic development agency, the Industrial Development Authority (IDA), has consciously concentrated on measures to attract hi-tech corporations – especially those of US origins – specializing in electronics, pharmaceuticals, chemicals and machinery. This policy of 'industrialization by invitation' (Jacobsen 1994) stresses Ireland's numerous attractions for foreign direct investment. These include access to the European market, significant corporate tax incentives, relatively low labour costs and the provision of special development zones complete with a variety of tax and capital grant concessions (see Sklair 1988 on the 'Shannon-Plassey package'). These factors along with the negotiation skills of the IDA have helped Europe's so-called 'Celtic Tiger' to attract 40 per cent of all American investment in European electronics since 1980 (*The Economist* 17.5.1997). The electronics sector, for example, now contains such major global names as IBM, Hewlett-Packard, Intel, Dell, Nortel, Gateway, Fujitsu and Motorola. Similarly, the pharmaceutical and chemicals sectors include such companies as GlaxoWellcome, Loctite, Merck, Sharpe and Dohme, Pfizer and Schering Plough. By the mid-1990s the Industrial Development Authority could report that MNCs accounted for almost half of all manufacturing employment (*The Irish Times* 25.6.1996) and 40 per cent of all exports. The electronics sector, which has the largest number of overseas employers, contained about 250 electronics companies, employed some 25,000 people, and generated over 25 per cent of the country's total exports (McGowan 1991: 43).

Multinational corporations, new technology and the expansion of technical education

Prior to the influx of MNCs, most of the scientific–technical labour in the Irish economy was absorbed by industries based on national resources and, from 1930, by public utilities such as electricity generation (OECD 1964). The arrival of hi-tech MNCs led to a widely reported skill shortage problem. These organizations required greater numbers of chemists, engineers and technicians than that previously available. Furthermore, they also needed employees who could adapt to what were effectively new technologies in the context of Irish industry. To address this problem the state embarked on a programme of expansion in technical education. Technical education was enlarged within the university system and a special tier of higher education colleges (Regional Technical Colleges) was created to provide an accredited system of technician education. Dramatic expansions also took place in science and engineering courses in the University sector. For instance, there was approximately a 100 per cent increase in the number of primary (98.9 per cent) and higher degrees (101.6 per cent) awarded in science between 1982 and 1990. During the same period there was an even greater increase (134.5 per cent) in the number of primary degrees awarded in engineering. Postgraduate degree awards increased by 40 per cent over the same period (all calculations derived from data contained in various reports of the Higher Education Authority).

This expansion in the supply of scientific–technical labour was, in turn, used to fuel the IDA's attempts to procure investment projects from US corporations looking for access to the European market. The ready supply of highly educated, but inexpensive, English-speaking graduates has, according to the IDA, become one of Ireland's unique attractions for mobile MNC projects (*The Economist* 17.5.1997; *The Irish Times* 25.6.1996). The MNCs themselves have also admitted how they were attracted by the availability of a young, highly educated workforce (*The Irish Times* 19.7.1996). In some celebrated cases, the IDA managed to win investment projects over its Scottish counterpart, the Scottish Development Authority, because Ireland could offer a more highly educated pool of graduates and secondary school students (*The Irish Times* 11.1.1991). More recently, it would appear that Ireland's unique ability to provide a relatively cheap supply of graduate labour has become so successful that the country is, according to the managing director of the IDA, attracting 'highly sophisticated projects at the leading edge of research and development' (McGowan 1991). The IDA has even claimed in its promotional literature that the restructuring of technical education has enabled it to attract so many leading leading edge hi-tech firms that Ireland had become 'the new hi-tech centre of Europe' (IDA 1984).

From these claims it would appear that some of the recent success of the Celtic Tiger can be attributed to the low cost and high quality of its skilled labour. Technical workers, for their part, would appear to have become a key group within the Irish economy since they appear to be critical in both attracting hi-tech firms and in contributing to their subsequent success. While it would be an exaggeration to say that Irish industrial policy is heavily dependent on its technical labour force, it would be reasonable to claim that these employees have achieved unusual importance in the nation's economic project.

However, by the late 1980s it had become apparent that the decision to expand technical education was 'based on a very naive understanding of the demand and supply of qualified labour' (Wickham 1992). The supply of engineers, scientists and technicians had increased to the point where it eventually outstripped the demand from industry, notably in the case of electronics technicians (EOLAS 1989). This development has generated a number of distinctive trends in the Irish labour market.

First, it has led to a situation where entry into a technical position is only possible after the successful completion of third-level education. This requirement means that access to skilled jobs in the chemistry and the micro-electronics-based industries is now rigidly linked to the possession of educational qualifications (Murray and Wickham 1987; McGovern 1995). The strength of this linkage is such that technical occupations have become exclusively qualification-based occupations. Second, the over-supply of technical labour has led to a 'brain drain' in which many Irish graduates find that emigration is the best route to professional employment. Between 1984 and 1988, for example, 43 per cent of those who had been awarded Ph.D.s in science emigrated compared with only 20 per cent of those awarded Ph.D.s in 1974–78 (*The Irish Times* 9.11.1990). The emigration of engineers was even more dramatic in the late 1980s. In 1987, for example, half (48.8 per cent) of those who graduated as engineers emigrated (HEA 1987). The brain drain is now so acute that Wickham (1992) in his review of these developments, argues that an engineering education has virtually become 'an education for emigration'.

How is this possible when Ireland has one of the lowest proportions of scientists and engineers in Europe (NESC 1985)? The answer is that the employment opportunities do not exist despite the IDA's success in attracting MNCs. While it should be noted that there were twice as many engineers employed in Ireland in 1989 (8,500) as compared with 1971 (*The Irish Times* 1.12.1989;

Cogan and O'Shea 1979) the general problem is that Irish industry simply does not require significant numbers of scientists, engineers or technicians. In relation to the chemical and pharmaceuticals industry, Ireland is seen to be too small to supply enough scientists for large-scale R&D activity. Moreover, the kind of MNCs that Ireland attracts would not appear to be those that undertake significant R&D activities.

Hi-tech Ireland and the new international division of labour

In contrast to the IDA's presentation of Ireland as the 'hi-tech centre of Europe', there is a body of research which indicates that this is more of an inspired illusion than a reality. The Telesis Consultancy Group (NESC 1982), which was invited in to review industrial policy in Ireland, found that foreign firms did not undertake any R&D activities. They concluded that:

> Foreign-owned industrial operations in Ireland with few exceptions do not embody the key competitive activities of the businesses in which they participate; do not employ significant numbers of skilled workers; and are not significantly integrated into traded and skilled sub-supply industries in Ireland.
>
> (NESC 1982: 193)

This situation has been interpreted as the result of dependent relations with more developed economies, (Long 1980; O'Hearn 1989; Jacobsen 1994), as 'late industrialization' (O'Malley 1985), and as the result of the emergence of the new international division of labour (Frobel *et al.* 1980; Henderson 1989) in which Ireland can be placed with other countries of the periphery (Perrons 1981; 1986). Common themes in this literature are Ireland's rigid reliance on free trade, its dependence on foreign companies for industrial modernization and, ultimately, how it has come to be exploited by 'footloose' MNCs in search of cheap labour. Both the proponents of the late industrialization and the new international division of labour (NIDL) perspectives highlight the low skill levels in MNCs and their lack of integration into the local economy. O'Malley (1987) concludes from his survey of the engineering industry that while more sophisticated products may be manufactured by these hi-tech firms 'typically only those *stages* of their production which make few demands on technical skills or local high-quality suppliers' are located in Ireland (1985: 149). He notes that while skill levels have been increasing the engineering industry still lacks firms that undertake advanced technical activities. Yet it is also clear from his research that MNCs are much more likely than indigenous firms to undertake some form of design or development work (O'Malley 1985; 1987; see also NESC 1982). Perrons claims that Ireland meets the needs of MNCs in the new international division of labour by simply offering a ready supply of relatively cheap labour. She summarizes the existing criticisms of Ireland's development strategy as follows:

> the low level of value added in Irish firms, the lack of linkages with local firms, the lack of research and development activities and the lack of development of new products in Ireland, meaning that the economy becomes dependent on imported technology with few skills being developed in Ireland.
>
> (Perrons 1981: 95)

In short, this literature indicates that the MNCs operating in Ireland have limited levels of technical activity, contain small proportions of professional and technical personnel and are characterized by relatively low skill levels. In this context, the expectation is that the chemical and pharmaceutical firms in Ireland will continue to make products that are based on research carried out at the parent company and will require finishing elsewhere (*The Irish Times* 24.4.1991). Although the theory of a new international division of labour can be criticized for its narrow view of the direction of technological change and its relationship to location (Henderson 1989; Schoenberger 1989) its description of the kind of activities which MNCs establish in the economic periphery of Europe remains unchallenged. Even so, one of the major weaknesses in the application of the NIDL thesis to Ireland is the lack of empirical evidence on the nature of employment within MNCs. Most studies simply refer to the findings of the Telesis Consultancy Group published in 1982 (e.g., Perrons 1981; 1986; Sklair 1988).

This study will seek to make a contribution to this question by examining the nature and role of technical employment within MNCs. The important issue here is the nature of the technical activities undertaken in such establishments. If these are of a routine nature and low in skill content then the employees may feel overqualified for the work that they are doing. To adopt the terms used by Jones *et al.* (1993), they may be best viewed as 'technical labourers' rather than as 'elite human resources'. This indeed was the claim made by Jones *et al.* (1993: 43) who, in a study of graduate engineers in nine British transnational firms, found that they were employed either as technical specialists or surrogate technicians. If this is the case then the work situation of technical professionals may not correspond with that implied by the service contract. The skill requirements of branch plants may be so low that these employers may be able to achieve control through task simplification and standardization. If this is the case then the employment relationship for technical staff need not be founded on a service contract.

MNCs and human resource management

Despite the supposed limitations in employment, there is a considerable body of evidence to show that it is also the branch plants of MNCs – especially those of American origin – that possess the most sophisticated approach to human resource management in Ireland. By 'sophistication' I mean the presence of a large range of formalized policies for employee recruitment, retention and development as well as employee relations matters generally. The largest existing national survey of employee relations reveals that when compared with Irish firms, foreign multinationals are more likely to have a personnel/HR function (with the Director having a position on the company board), to operate merit-based payment systems, to make greater use of internal labour markets and to have a more coordinated and planned approach to employee relations management (Gunnigle *et al.* 1994). Of particular interest to this study is the finding that the presence of performance-related pay in US companies is almost twice the mean average of Irish companies (Turner *et al.* 1997). Earlier research by Murray *et al.* (1984) (albeit of a smaller national sample) found that US and, to a lesser extent other foreign firms, were more likely to have written personnel policies and more sophisticated policies on remuneration, information disclosure, communications, employee development and recruitment. Kelly and Brannick (1985), also concluded that US-owned subsidiaries implemented more elaborate personnel management practices than indigenous employers. For instance, they undertook careful employee selection procedures, gave responsibility to line management for the implementation of human resource policies, had some form of employee consultation ranging from suggestion schemes to formal works councils and had explicit statements of their human resource philosophy.

With regard to union recognition, Turner *et al.* (1997) find that there is little difference between Irish and US companies. Though a slightly higher proportion of Irish companies recognize trade unions, the difference is not statistically significant. There is, however, evidence to show that since the 1980s some US MNCs in the electronics sector have developed a viable strategy of 'union substitution'. These firms adopt a range of measures that are deliberately designed to pre-empt the need for trade union representation (Toner 1987; Flood and Toner 1997). Many of these MNCs began production on greenfield sites (Gunnigle 1992), operate in uncertain and competitive international product markets (Flood *et al.* 1994; Roche and Geary 1996) and attach greater priority to human resource matters (Turner *et al.* 1997). Their overall approach to labour management tends to be based on individualist rather than collectivist principles of employee relations (Gunnigle *et al.* 1997). Consequently, it is not surprising to find that the arrival of non-union hi-tech MNCs represents a serious challenge to the trade unions' ability to win recognition from employers for the purposes of collective bargaining (McGovern 1989).

Thus, employment within multinationals would seem to represent something of a paradox: their structures and practices can be accommodated within such diverse models as those of the NIDL and of 'best practice' HRM. They are typically low-skill employers yet they provide a broader range of formalized HRM policies than Irish-owned firms. At the same time, they do not appear to differ greatly from Irish firms in terms of conformity with the major institutions of industrial relations (Kelly and Brannick 1985; Turner *et al.* 1997) yet they seem to place greater emphasis on developing an 'individualistic' relationship with their employees than on their dealings with trade unions (Roche and Geary 1996). This seeming paradox can be resolved once it is appreciated that these MNC employers offer superior employment conditions than those of indigenous Irish employers. This is clear from the literature on human resource management and in research on job security where direct comparisons are made. These indicate that, contrary to the prevailing myth of the 'footloose' foreign firm, employment in MNCs, especially those of US origin, tends to be better paid (Toner 1987) and more secure than in Irish-owned firms (McAleese and Counahan 1979; EOLAS 1989; Keating and Keane 1991).

Comparisons of skill levels are not really possible because of the lack of matching Irish-owned firms. This is perhaps the most revealing difference between the indigenous and foreign-owned enterprises in Ireland's economy. It also explains why so many Irish engineers and scientists are forced to emigrate to find employment.

Research methodology

HRM practices of the kind described above are one of the key elements in the argument about changes in the employment of the service class. It is therefore essential that this research is based on firms that are at the 'leading edge' of human resource management. The research strategy is therefore based on that of the *critical case* approach (Goldthorpe *et al.* 1968). That is to say, it is based on the idea that the phenomenon which is being studied is more likely to be found in situations that meet certain theoretical requirements rather than elsewhere. The *locales* that are chosen should therefore be as favourable as possible for testing the hypothesis in question. To paraphrase Mitchell (1983), the cases should be selected for their explanatory power rather than their typicality; for their theoretical potential rather than their representativeness. The underlying approach is therefore Popperian in that a specific thesis is being tested on the basis that it can be falsified (Popper 1959). This was one of the reasons for choosing MNCs in the Irish context. The other was that hi-tech multinational firms were much more likely to employ a *critical mass* of technical workers.

Given the focus of this research, case studies were chosen as the best means of examining employer practices in detail. Crompton and Jones (1988) claim, among other things, that this method is particularly useful in investigating promotion patterns and internal labour markets in the white-collar context because it enables the researcher to penetrate the often confusing world of organizational job titles. Whatever the merits of single case studies I believe that these are of limited value in comparison with a multiple case approach that allows for the replication of single case studies on a systematic basis, for comparisons between cases and for the elaboration of different aspects of the research question. This multiple case approach, as Eisenhardt (1989; 1991) argues, shifts the balance of case study research from that of 'story telling' to theory generation.

Finally, I would like to make it clear that this study does not set out to describe a general historical trend towards the erosion of the service contract. That would require a longitudinal study of specific variables. What this case study research does instead is provide a detailed examination of the impact of the employment practices that are supposed to drive this trend. In doing so, it aims to assess the accuracy of the service contract conception of employment in the context of HRM and other management practices.

The branch plants

Using a database provided by EOLAS, The Irish Science and Technology Agency, a total of twenty-two firms from the electronics and chemicals sectors were contacted. Those selected were judged to be the most likely employers of large numbers of technical professionals. This judgement was based on information drawn from the EOLAS database, an extensive review of newspaper and magazine articles on various companies and from information supplied informally by individuals employed in these industries. Of those firms who responded, six agreed to participate in the research. Funding constraints limited the fieldwork to four of these organizations. The final group comprised two pairs of 'matched' firms from pharmaceutical/chemicals and from electronics (Figure 2.1). All had an R&D or Technology unit in addition to a manufacturing facility; were branch plants of North American corporations; manufactured for export markets (primarily Europe); and implemented a comprehensive range of HRM policies. They also belonged to the *Fortune* 500 table of the largest corporations in America.

Their size and scale of operations is such that they could bear the fashionable labels of 'transnational' or 'global corporations' (see Sklair 1998). I have chosen to use the term 'multinational' simply because there is less uncertainty over its meaning. This will not be

Industry	**Pharmaceuticals/chemicals**		**Electronics**	
Firms	**Kenine**	**Harford**	**Mertel**	**Trojan**
Employees	270	400	350	500
Technical staff %	14.8	22.5	22.8	12.0
Occupations	Technicians Chemists Chemical engineers	Technicians Chemists Chemical engineers	Technicians Engineers	Technicians Engineers
Union members	Technicians	Technicians	None	None

Figure 2.1 The MNC branch plants

obvious from the names used here since one of the conditions of gaining access into these firms was an agreement to use fictitious names. One difference worth noting is that the technicians were unionized only in the two chemistry-based firms. None of the chemists or engineers were formally represented by trade unions.

Kenine Pharmaceuticals

Kenine Pharmaceuticals has been repeatedly voted America's most admired company in the annual *Fortune* magazine survey. At the time of the research it employed some 34,000 people in eighteen different countries around the world. The Irish branch of Kenine Pharmaceuticals, which employed 270 people, was located in a scenic rural area, ten miles outside the town of Clonmel in the South-East of Ireland. A subsidiary of Kenine's manufacturing division, it manufactured a range of human health products for the treatment of arthritis, high blood pressure, high cholesterol and depression. These products were then exported in bulk form to the European Marketing division of the corporation where they were formulated into final dosage form before being packaged and sold. Accordingly, the Irish factory did not deal directly with customers outside the corporation itself. Its customers were other divisions of the Kenine corporation. The R&D division of Kenine Pharmaceuticals was located in New Jersey, USA.

In 1985 a new development laboratory was built at the Irish site to provide technical support for European manufacturing operations within the corporation. Most of the forty technical staff worked in this new department called technical operations. This purpose-built facility, which had offices for chemists and chemical engineers, stood to one side of the site some distance from the main production plant itself.

Some manufacturing engineers had offices in a neighbouring building, but most of the remaining technical staff worked in the quality control department. The quality control laboratories were housed in the main production building. On establishment the firm entered into a closed shop agreement with the Irish Transport and General Workers Union (now Services Industrial Professional and Technical Union – SIPTU) for its general operatives and laboratory technicians. The laboratory technicians were included in this closed shop arrangement since it was generally accepted that technicians were union members at that time. Prior to the arrival of large numbers of MNCs, most technicians were employed by the state in national testing centres or small laboratories. Since these were all traditional trade union strongholds, management accepted the prevailing labour market norm that technicians were union members. In Kenine, the technicians elected one shop steward to the workplace committee. Similar arrangements existed for the craftworkers who were divided into three unions, the Amalgamated Engineering Union (AEU), the Electrical Trades Union and the National Engineering and Electrical Trade Union (NEETU). Though I was informed by my 'gatekeeper', the personnel manager, that although there had been recent industrial relations problems he was proud of the fact there had never been a strike at the Irish plant.

Harford Laboratories

Harford Laboratories set up a subsidiary plant in Dublin in 1966 in order to expand its operations into Europe. It also took the unusual step of basing part of its research group in Ireland. By 1971 this group consisted of twelve full-time laboratory staff. After a number of early technical breakthroughs the R&D facility grew until it had seventy R&D staff, ten individual laboratories and a pilot plant by the time of this research. It was at that point, and still is, the largest private sector-research group in Ireland.

The manufacturing facility also expanded in the 1980s with the addition of a second production plant alongside the new R&D laboratories at a site some six miles away from the original plant. Harford was (and is) the largest adhesives manufacturer in Ireland, employing some 400 people. The Dublin plant, which was the European Manufacturing and R&D headquarters, developed and manufactured a range of anaerobic and cyanoacrylate adhesives that were exported to European, Middle Eastern, Far Eastern and African markets. These were sold to two types of markets, individual consumers and to other industries with the automobile industry being one of its largest customers. One of its adhesive's products, which had been refined at the Dublin R&D site, was listed in *Fortune* magazine in 1979 as being among the business triumphs of the 1970s.

Most of the chemists, chemical engineers and technicians were employed in the R&D unit that had a separate building and entrance on the new site. The R&D unit is described as being very different from manufacturing; managers occasionally went over there for 'peace and quiet'. The rest of the technicians worked in the quality control departments in both the old and new manufacturing facilities. Most of the younger technicians worked at the new site either in R&D or in quality control.

When Harford Laboratories initially started up in Ireland it did not have a specific personnel management function. Its employees had begun to organize by the early 1970s with the general employees joining what was then the ITGWU, now SIPTU. The technicians started to unionize in 1970 with what was then the Draughtsmen and Allied Trades Association (DATA), a British-based union. By the time of the research this trade union had evolved into the Manufacturing, Science and Finance Union (MSF). In Harford the R&D technicians elected two representatives while the quality control technicians and the clerical staff elected one each. On my first visit to the plant the technicians' union was described, by a member of the personnel department, as being 'active rather than militant'. The craft workers were represented by the NEETU.

The Harford management responded to the increasing industrial relations workload that followed the emergence of trade unions within the firm by creating a personnel department in 1977. The first agreement on rates of pay, conditions and work duties was agreed with each union in the same year. Unlike the other companies, there have been a number of short strikes in Harford. The initial starting date for the research in this company was delayed for five months because the management were eager to have the negotiations for the next Comprehensive Agreement finished before the research began. After the research commenced, my presence was initially treated with suspicion by some MSF members before I convinced them I was not a management spy!

Mertel Telecommunications

Mertel Telecommunications is one of the largest telecommunications firms in the world employing approximately 47,500 people in fifty-five manufacturing plants in Canada (country of origin), the US, England, Turkey, Malaysia, Brazil and Ireland. The Irish factory, which initially employed thirty people in the manufacture of rotary dial telephones was set up in 1973 outside Galway, in the West of Ireland. By the early 1990s Mertel designed, developed and manufactured a large range of telecommunications equipment including telephone sets, digital Personal Automated Branch eXchanges (PABXs) and data switching devices. These products were exported directly to European, Middle Eastern and Far Eastern markets.

The Irish plant employed some 350 employees of which eighty were either engineers, technical specialists or technicians. Of these about fifty-four engineers and twelve technicians worked in the technology unit attached to the factory. This unit shared in the design and development responsibilities for the European market. The basic design and development activities were carried out in the corporate research centre located in Canada. Engineers were also employed within the Irish plant in the production and quality control departments. The remainder of the technicians worked in quality control and production.

Like Kenine Pharmaceuticals, this company also commenced production with a closed shop or 'sweetheart deal' with the Federated Workers Union of Ireland (now SIPTU) for its production staff. The salaried staff, which includes engineers and technicians, are not allowed to join this or any other union. The wages of the SIPTU members were set through annual negotiations between the company and the union. There has never been an official strike in the Irish plant.

One of my first impressions on touring the plant was of the disproportionate number of women working in semi-skilled assembly jobs. This, however, was a phenomenon that existed from the day it commenced production. A press release on Mertel's arrival in Ireland reported that it would employ 500 people by 1977 of whom only 40 per cent would be male. The busy atmosphere and the pressures of time were used by some managers as an argument to limit my time and number of interviews with each department. Though similar points were made by the Director of the technology unit, the work environment of this group was considered to be very different from that of the main plant. Their working environment was thought to be quieter, less stressing and generally more pleasant especially as the engineers had much more freedom to decide how they would spend their time without having it decided for them by the demands of production schedules. In the words of the personnel officer: 'They're looked upon as being in heaven'.

Trojan Electronics

Trojan Electronics was founded in the USA in 1951 as a specialist electronic laboratory. The expertise that it built up at its R&D centre – located in Massachusetts – was a major contributor to the company's early success. It pioneered the desk-top calculator and, in 1972, launched one of the earliest word-processing systems. Trojan expanded dramatically with the rise of the market for Personal Computers (PCs). By the time it arrived in Ireland in 1980 it had sales and service operations in 127 countries around the world. It was located in a university science park outside Limerick, in the South-West of Ireland. Trojan Electronics manufactured a range of Vertical Storage (VS) CPUs (mainframe computers) and PCs for European, African and Middle Eastern markets. The manufacturing facility was supported by three departments: manufacturing engineering, test engineering and quality control all of whom employed engineers.

Like other Trojan electronics factories, the Irish plant was set up on a non-union basis. The company's espoused philosophy was not that it was anti-union, merely pro-employee, giving special attention to conditions of work, canteen and recreation facilities. All employee relations problems were expected to be resolved internally through the firm's open door policy. Following this line, the members of the personnel department were called 'personnel area representatives' and each had special responsibilities for particular categories of workers. The personnel area representative for the engineers and technicians was the initial contact within the plant. There had never been any widespread attempt at unionization by employees nor had there ever been a strike in this plant.

At one point in its history Trojan Electronics employed over 700 employees in its Irish plant of whom approximately ninety were employed as technical staff. At the time of the research this had dropped to 500 of whom sixty-five were employed in a technical capacity. Some five months previous to this, the corporation had announced job cuts at its operations around the world because of continuing financial losses. In 1988 it had a turnover of $3,068 billion but losses up to the last quarter of that year amounted to $54 million. Trojan Electronics had already closed a Scottish facility and transferred its operations to the Irish plant in the previous year.

My first visit to this firm followed on the implementation of a redundancy programme that had laid off 120 white-collar and technical staff ('indirect labour'). The white-collar areas were targeted as part of a corporate wide 'delayering exercise' that aimed to reduce the layers of management between the corporate Director and shop-floor employees. During the course of the research, a series of speculative rumours swept through the plant to the effect that the company would either be subject to a takeover or closed or possibly even both. There was, understandably, an edge of nervousness in the plant about its future. The technical staff had been worried about the increasing levels of routine in their work as a result of the company's failure to introduce any new products during the previous three years. A particular concern was the future of indirect labour within the plant. A new product line was expected to arrive in the Irish plant at the time of this research but it was delayed by lay-offs at the US-based corporate R&D centre.

Technical workers

Before proceeding it is necessary to clarify some of the key terms employed in this study. That of 'technical workers' is of major importance since I shall use it and other generic titles such as 'technical labour' to cover a number of different technical occupations. Smith (1987: 1–2) points out that such terms are of limited value since they refer to a heterogeneous grouping of third-level graduates, the professionally qualified along with specialist manual and maintenance workers. To overcome this problem I shall use terms like 'technical workers' when referring to engineers, chemists and technicians. In doing so, I draw upon Barley and Orr's (1997) 'trait' definition of technical work since it allows professionals and technicians to be so categorized. Their definition comprises:

(a) the centrality of complex technology to the work,
(b) the importance of contextual knowledge and skill,
(c) the importance of theories or abstract representations of phenomena, and
(d) the existence of a community of practice that serves as a distributed repository for knowledge of relevance to practitioners. (1997: 12)

In addition, I shall show (in Chapter 3) that, in the Irish context, the term 'technical workers' refers to individuals who possess third-level educational qualifications.

In the context of the literature on internal labour markets, the conceptual distinctions made by Althauser and Kalleberg (1981) are useful in locating and describing the labour market position of various technical workers. Althauser and Kalleberg distinguish between firm internal labour markets (FILMs) that are established by, and confined to, a single employer and occupational internal labour markets (OILMs) that exist for a particular occupation, and are not necessarily restricted to a single employer. Engineers and chemists, who typically possess a high degree of externally acquired knowledge that is often of a theoretical nature, occupy OILM jobs. Like those in FILMS, 'on-the-job' training facilitates movement up the job ladder. In contrast to FILMs, the progressive development of skills and knowledge in OILMs depends on the possession of external qualifications. These also act to withstand skill obsolescence as a result of new advances in technical knowledge.

While these characteristics obviously apply to the jobs occupied by the engineers and chemists, technicians are located in what Althauser and Kalleberg (1981) describe as an occupational labour market (OLM). In comparison with those in OILMs, technicians do not have quite the same level of education and are likely to be involved in more specialized activities, a point that I confirm in Chapter 4.

Finally, it is worth noting that while 'higher grade technicians' are members of the service class, 'lower grade technicians' belong to an 'intermediate group' (Class V) along with supervisors (Erikson and Goldthorpe 1993: 38–39). The work of such technicians has a significant manual component though they have relatively high levels of remuneration and reasonable job security (Goldthorpe 1980: 41). I have incorporated the technicians into a study of 'service class' employment for a couple of reasons. First of all, elements of their work and market situation have features of the service contract, or, of Fox's (1974) 'high-trust high-discretion' work roles. Second, given these comparable elements I wished to understand how technicians may be differentiated from professional chemists and engineers especially when they work so closely together within organizations.

Research methods

The research methodology was influenced by that used by Lincoln and Kalleberg (1990) in their cross-national research on work organization, culture and commitment in the US and Japan. In particular, their organization-based two-stage case study design along with their combination of detailed interviews and questionnaire data suggested an exemplary model for organizational research. Consequently, four different methods of data collection were used in this study:

1 semi-structured interviews with managers, supervisors, chemists or engineers, technicians and shop stewards;
2 periods of non-participant observation of technical work;
3 documentary analysis of company records and policies, third-party arbitration records and surveys of the labour market for science and engineering graduates; and
4 a self-administered questionnaire of the technical workers in all four factories.

This research therefore goes further than existing work on the management of technical work, notably that of Causer and Jones (1990: 4; 1996) which only draws upon interviews with personnel managers and technical managers. Most of the material reported here draws on the interviews and the survey data.

I conducted a total of seventy-nine interviews with managers, engineers, chemists and technicians during the first phase of the research (see Appendix for details). These were used to obtain information on specific areas of the management of technical staff and, also, to help prepare the questionnaire survey used in the second phase of the research. The interview material was sorted and categorized by using the ETHNOGRAPH software package for the analysis of text-based data (see Fielding and Lee 1991 for a discussion of these methods). During the second phase of the research (July–August 1991) 203 questionnaires were distributed across the four plants to engineers, chemists, chemical engineers and technicians. One hundred and sixteen responded giving an overall response rate of 57 per cent.

A notable feature of the technical staff covered in the questionnaire survey was their relative youthfulness: almost two-thirds (64.6 per cent) of the respondents were less than 30 years of age. However, more than one-third (37.9 per cent) of the respondents were in the 31–40 age group and 17.2 per cent in the 41–52 age group in Harford Laboratories – the longest established firm. One implication of the age distribution of the overall sample is that many of these technical workers were still in the formative years of their careers. As a result, a significant proportion, as I describe in Chapter 7, were prepared to leave in order to advance their careers as professional engineers, chemists or technicians.

In gender terms, these occupations were overwhelmingly male-dominated (81.9 per cent). Less than one-fifth were female (18.1 per cent) and of these almost two-thirds (63.7 per cent) were employed as technicians while the remainder worked as chemists

Table 2.1 Respondent's occupation by firm

Company	*Kenine* %	*Harford* %	*Mertel* %	*Trojan* %	*Total* %
Technician	5.6	48.4	40.0	67.6	45.7 (53)
Engineer	38.9	9.7	60.0	32.4	34.5 (40)
Chemist	55.5	41.9	0.0	0.0	19.8 (23)
Total	100.0	100.0	100.0	100.0	100.0
N=	18	31	30	37	116

or engineers. Details of the occupations and numbers of the respondents from each company are listed in Table 2.1.

Finally, it should be obvious that generalizations from these cases cannot be extended to MNC branch plants in general on the basis of statistical representativeness. However, since this research is based on 'strategic cases' that were chosen on theoretical grounds, it can be argued that its findings are of general significance in two respects. The first is that if the theories of service-class employment are losing their explanatory powers then it should be evident in those organizations that have the most advanced HRM techniques. If this is not present in the firms studied here, then it is unlikely to be present anywhere else. The second claim relates to aspects of the new international division of labour thesis. If there are not fully fledged R&D activities and concomitant technical career ladders in these firms, then they are unlikely to be present in the Irish private sector generally since these were selected precisely because they represent the technical elite. Alternatively, if they are, it would be difficult to make any wider generalizations because of the select nature of these firms.

3

RECRUITMENT, OCCUPATIONS AND QUALIFICATIONS

Introduction

One of the characteristic features of the market situation of service-class employment has been the prospect of a career along a well-defined path. The understanding was that employers offered secure employment and the prospect of regular promotion in exchange for a high degree of employee commitment. The existence of this implicit contract is so well established in the literature on professional and managerial employment (e.g., Kanter 1977; Rosenbaum 1984; Nicholson and West 1988) that it has come to represent the archetype against which various forms of change are assessed. In this regard much of the recent literature on the impact of organizational downsizing has been concerned with the question of whether the traditional career model has been replaced by a new 'professional' career model (e.g., Heckscher 1995; Kanter 1989). For our purposes, the most relevant conclusions from this literature are, first, that organizations are choosing markets over hierarchies as a means of structuring their staffing requirements and, second, that professionals have a much greater 'market capacity' (Giddens 1973: 103) than managers by virtue of their possession of academic qualifications (Savage *et al.* 1988; 1992).

This chapter is the first of two to examine employers' labour market practices in detail. The central question (here and in Chapter 7) is whether employers continue to rely on the internal job hierarchies or have instead moved towards market solutions. In other words, the issue is one of a choice between internal versus external labour markets. If employers are indeed changing the service contract then the Irish labour market is a particularly appropriate testing site since it is marked by an over-supply of graduate technical labour. This situation contrasts dramatically with the situation in the UK (Winstanley 1991; Causer and Jones 1993) and in the US (Turbin and Rosse 1990) where there has been a continuing problem of skill shortages. There are two parts to this question. The first part, which relates to employers' retention practices, asks if the specialist knowledge and expertise of these employees is such that hi-tech employers continue with internal hierarchies regardless of labour market conditions. This issue of employee retention will be taken up in Chapter 7.

The second relates to the links which employers make between levels of educational qualification and work roles. It might be expected that in such propitious labour market conditions as those described here employers would simply hire the most qualified candidates. Dore's (1976) thesis about the relationship between national patterns of education, qualifications and development is of relevance here. He claims that, in countries with a relatively late experience of industrialization, such as Ireland (Goldthorpe and Whelan 1992), the educational qualifications of those holding jobs seem to be influenced more by the supply of qualifications than by the needs of the actual jobs. One symptom of the 'late development effect' is what he terms 'qualificationism': a situation where educational qualifications become a prerequisite for those seeking employment. In some cases, a surplus of labour forces graduates tend to apply for jobs that are one level below what they would have conventionally obtained. Dore argues that in such situations employers automatically select the most qualified because they believe that the more highly qualified candidates will be of greater use to them as employees (see also Collins 1979).

Since the employers studied here are in electronics and chemistry-based industries it might be assumed that an increase in the qualification levels associated with specific occupations would be of particular value to them. There are a number of reasons for this. First, hi-tech firms generally tend to employ a relatively large proportion of engineers and scientists in comparison with more traditional industries (Anderson and Kleingartner 1987). Second, there is a positive relationship between the competitiveness of 'hi-tech' science-based firms and the technical knowledge and expertise of their staff (Bosworth *et al.* 1992). This means that these employers not only have the opportunity, they also have the incentive to erode one of the key features of professional employment. Furthermore, if employers were to recruit directly from higher education they would be free to allocate individuals to positions as they would not be dealing with employees who had formed a strong sense of occupational status or identity on the basis of previous employment experience. The question I examine is whether these employers actually exploited the prevailing labour market conditions by selecting the most qualified candidates available for any given position. By doing so, they would have undermined the market situation of professional employees who, according to Savage *et al.* (1992), have traditionally been able to rely on their educational qualifications to secure their position.

This chapter will begin with a brief description of labour market sourcing practices before moving on to the subject of qualification inflation. The remainder of the chapter examines the 'social' criteria that managers use to select new staff. These criteria are important because they reveal much about the labour market dominance of these firms as well as the 'right attitude' for service-class employment.

Employer sourcing practices

When these four MNCs first arrived in Ireland (mid-1960s to mid-1970s) they found it difficult to recruit technical staff because the hi-tech labour market was quite small. Irish students seemed to have little interest in science and technology and the universities only offered a small number of places. None of the Regional Technical Colleges (RTCs) or National Institutes of Higher Education (now technological universities) existed at that point. Trojan Electronics, for example, was forced to set up its own 'in-house' technician training programme because it could not find skilled employees with experience of electronics – not to mention qualified technicians. Some of its initial 'technicians' were converted marine radar and radio operators. However, the state-initiated expansion of technical education, beginning in the late 1970s, had transformed the recruitment experience of these organizations. According to the personnel manager in Harford Laboratories, the quality and the quantity of technicians had improved dramatically since then. In the mid-1970s, his firm had struggled to find technicians with basic City and Guilds qualifications (Part II). By the mid-1980s the situation had changed so much that they would only recruit RTC diploma students who had graduated near the top of their class. These candidates were, in his words, 'superb' because they had experience of instrumentation that did not even exist in the 1970s. These changes also made it much easier to recruit technical staff. The labour supply was such that his department was regularly 'showered' with speculative job applications. From his perspective it was a 'buyer's market', although one that was based on inadequate educational planning by the state: 'Frankly a simple advert in a newspaper will get hundreds. It's appalling I might add from an educational point of view but nevertheless from Harford's point of view it is a buyers' market' (personnel manager, Harford Laboratories). Similarly the electronics firms also found it relatively easy to recruit technicians and engineers. Technicians were, according to the personnel officer in Mertel Telecommunications, 'spewing out' from the local RTC. With respect to engineers, the personnel area representative in Trojan Electronics stated that they never 'had any problems getting the graduate in, as in the raw graduate'. They did, however, experience some difficulty in obtaining specialist expertise such as 'vendor engineers' (who specialize in sourcing electronics components from industrial suppliers).

These MNCs considered it essential to recruit what they perceived to be 'the best' available technical talent. In Kenine Pharmaceuticals, they insisted that they were compelled to do so because of the constant competitive need to be more innovative. The future of the plant depended on its ability to compete, not only with those of other corporations, but also with 'sister' plants in other parts of Europe. Similarly, managers in Harford Laboratories claimed that they had to hire the best because their competitive edge was based on an ability to develop new products with commercial potential. For this reason these firms used a multi-method approach to trawl through the national labour market for the 'best'. According to the employee survey, most were either recruited through newspaper or magazine advertisements (30.2 per cent), college or university careers' offices (21.6 per cent) or through direct applications to the company (20.7 per cent). While no individual method stood out across all four firms, Harford appeared to have had most success with advertisements since over half its technical staff (61.6 per cent) were recruited using this method. Similarly, half (56.0 per cent) of the Kenine staff were recruited through university careers' offices. The electronics firms had approximately equal success with all three methods as none of them accounted for more than one-third of all appointments.

Although these employers used a variety of sourcing methods, they had all also begun to foster links with various universities and colleges in the Republic and in Northern Ireland. Some of these were formed through contacts with professors and lecturers who had taught the managers when they themselves were students of engineering and chemistry. In addition, all four firms provided industrial placements for students, hosted on-site visits and availed themselves of opportunities to give presentations on the company and its work. The industrial placement schemes were themselves a source of new staff. Those who had impressed would be invited to apply for jobs in the following year. Kenine Pharmaceuticals and Harford Laboratories also funded research projects while Trojan Electronics had deliberately located itself in a university science park.

These links were made with the intention of improving their recruitment process by creating a positive public image among students and their teachers. The reality, however, was that they enabled these employers to 'track the talent' among the final year students: 'We know always who the good people are in the colleges and we do that by involvement with the third years and by student placement programmes and so on' (Recruitment and training officer, Kenine Pharmaceuticals). As this manager indicates, they could obtain 'inside' information on how students from the same institutions compared with each other. In addition, their involvement with different colleges also helped to make judgements about how students from different institutions compared. Finally, the relationships they had developed with various university departments enabled them to check exam results with teaching staff, obtain reliable character references, and learn of students undertaking projects or dissertations relevant to the firm's activities.

Selecting 'the best': suitability and acceptability

To analyse the recruitment criteria used by these employers, I have drawn on Jenkins' (1982; 1988) well-known distinction between *suitability* and *acceptability*. Both of these terms refer to the types of criteria which employers use to determine the potential contribution of job candidates. Suitability is defined as being functionally (or job) specific. It typically includes formal educational, trade or professional qualifications. Acceptability criteria, which are functionally non-specific, are concerned with the management and control of work. These, according to Jenkins, are highly subjective and depend on managerial perceptions of such things as the job applicant's manner, attitude, origins and work history. From interviews with managers in these firms it would appear that candidates were chosen on the basis of their 'acceptability' having first demonstrated their 'suitability'.

Suitability: the erosion of conventional occupational standards?

From the literature it would appear that academic qualifications provide professionals with relatively secure labour market assets. The Irish labour market situation provided an opportunity to assess the strength of this claim in the context of possible trends towards qualification inflation. Two distinct types of qualification inflation were investigated. The first was where employers break with the previous links between qualifications and positions on their internal technical hierarchy. In this case, standards may be adjusted *across* the previously established boundaries of technical occupations. Traditionally, an NCEA certificate, or City and Guilds Part II, had been accepted as a technician qualification since the early 1970s. If employers had broken these links then they were likely to recruit those with degrees in engineering and science as technicians, that is, those who would previously have been employed as engineers and scientists. If this holds then there is evidence of inflation *across* technical occupations.

The second examines the case of an escalation of standards *within* individual technical occupations. In this case employers recruit the most qualified available for a position but without breaking the link between qualifications and occupations. This practice would inevitably lead to an escalation of standards within particular occupations. In the case of technicians, the emphasis would now be on higher level diplomas rather than on certificates or City and Guilds. In the case of engineers and chemists for example, employers might be expected to have recruited those with the highest level of postgraduate qualification regardless of the content of the position being offered.

In relation to the first question, the evidence from the questionnaire survey indicated that these elite employers had not adopted a general practice of employing those with degrees as technicians (Table 3.1). Most of those employed as technicians held 'subdegree' qualifications: diplomas or certificates (NCEA) or City and Guilds qualifications (88.6 per cent). Only 5.7 per cent of the technicians possessed degrees. Furthermore, interviews with technicians revealed that some of these qualifications had been obtained only after taking up employment. Most of the engineers (60 per cent) had degrees, while most of the chemists (60.9 per cent) actually possessed postgraduate degrees. This latter finding was explained by the large number of Ph.D. chemists employed in Harford's R&D unit.

Table 3.1 Occupations and qualifications

Occupation	*Technician* %	*Engineer* %	*Chemist* %	*Total* %
Postgraduate	0.0	15.0	60.9	17.5
Degree	5.7	60.0	21.7	28.1
Sub-degree	88.6	22.5	17.4	52.6
Other	1.9	2.5	0.0	1.8
No response	3.8	0.0	0.0	0.0
Total	100.0	100.0	100.0	100.0
N=	53	40	23	116

Note: Postgraduate = Ph.D., MSc, MEng, MBA; Degree = BSc, BE, BA, MRSC, ACCA; Sub-degree = Diploma, Cert., City and Guilds

This pattern did not emerge by accident. It was a direct reflection of the recruitment criteria adopted by these organizations. For example, the personnel manager in Mertel Telecommunications stated that they were quite strict about 'keeping diplomas for technicians and degrees for engineers'. This policy, which eliminates the possibility of inflation across occupations, was shared by each of the four organizations. The reason for this convention was brought out most clearly in an interview with a manager in Trojan Electronics. He outlined the problems that could arise in dealing with the expectations of 'engineers' who had been hired as technicians:

> 'You'd hire someone with a degree as a technician and within a year or whatever it takes they'd be looking to be made . . . "Well look I have a degree, I should be an engineer" . . . So you are creating a problem for yourself down the road. Basically we kept the degrees to engineers and certs and diplomas for technicians.'
>
> (personnel area representative, Trojan Electronics)

This policy also suggests that employers should not be viewed simply as passive beneficiaries of rising labour market standards (e.g., Oxenham 1984). The labour market practices of these MNCs were constrained by the way that their own previous practices had shaped employee expectations. Any attempt to break with the past would have ruptured the career expectations of newly recruited staff. It would also have led to employee resistance and to an inevitable erosion of morale among the technical staff.

Management obviously considered the costs of creating such a problem to be too high, at least at the time of this research.

There was, however, evidence of qualification inflation within the occupations of technician and chemist. In this instance, employers were quite prepared to reap the benefits of the increased supply of highly qualified technical labour that flowed into the Irish labour market during the course of the 1980s and 1990s.

Technician qualifications

Each of these firms had, as indicated earlier, experienced considerable difficulty in recruiting technicians when they first arrived in Ireland. Since then, however, the supply of technicians increased dramatically to the point where an over-supply of technicians was reported in a national survey of electronics firms in 1989 (EOLAS 1989). This trend was acknowledged by managers within each of the four firms. The dramatic increase in the supply of those with technician-level qualifications allowed them to be much more selective, thus contributing to an escalation in the level of qualification required for the role of technician. Prior to this increase, these employers had readily employed individuals with City and Guilds qualifications, or national certificates. Both of these qualifications could be obtained on the completion of two-year courses. Now they were in a position where they could insist that candidates had to have a national diploma which could only be obtained after three years within a higher education college (see also Murray and Wickham 1987). The following comments illustrate this point:

> 'For a technician . . . we start with diploma people.'
> (quality control supervisor, Kenine Pharmaceuticals)

> 'what you're looking at is to get the best. In technicians' terms that now means you can get the very best with an NCEA Diploma, not the Certificate, not a City and Guilds, not a Trainee Technician.'
> (personnel manager, Harford Laboratories)

The decision to only choose diplomas, which was facilitated by the surplus of technicians in the late 1980s, did, however, have negative implications for the position of technicians within the internal labour market (see Chapter 7).

Engineers and chemists

Strictly speaking, there was no evidence of qualification inflation in the case of engineers and chemists. However, standards had clearly risen. In the case of engineering, this trend can be at least partially explained by the limited supply of engineers with postgraduate qualifications. One of the reasons for this was possibly the strong demand for those with undergraduate degrees in engineering, as reported by the Association of Graduate Career Services in Ireland (1989). The two chemistry-based employers, however, had adopted a policy of only appointing those who held doctorates for the position of chemist: 'We're really, on the chemists side, only recruiting Ph.D.s now' (recruiting and training officer, Kenine Pharmaceuticals). This policy was driven by changes in the work of these firms rather than by changes in the labour supply. Two of these firms were beginning to conduct more sophisticated R&D activities with the result that they needed more specialized staff. In the case of Harford Laboratories, this policy had existed from the late 1970s when the firm had produced an extremely successful adhesive product. The subsequent expansion of the R&D unit led to the on-going recruitment of research chemists with Ph.D.s. This in turn contributed to the growing technical reputation of the plant.

The decision to expand R&D work in Kenine Pharmaceuticals had been, by contrast, a more recent one. The role of its technical operations unit was changed from one which merely provided technical support to the Irish plant to one that undertook research of a more long-term nature, commissioned by the corporate research group. Management were, as a result, eager to recruit individuals with the appropriate research training and expertise to undertake the new technical activities. In the words of the personnel manager, it made sense to hire the most qualified candidates because of 'the more creative nature of this work'.

Finally, it was apparent from the interviews that experience of this kind of work was obviously important. Engineers and chemists who could take on projects after a few months with the company were highly valued. For technicians in the chemistry-based firms, experience of particular types of equipment was valuable since their work tended to be dominated by the use of specialist analytical equipment. The interesting point here, however, was that these employers did not always select those candidates who had most experience. While some amount of experience was desirable, since it indicated that the applicant could be trusted with expensive equipment, too much experience meant that the individual would have to be 'reprogrammed' to fit the company's own work procedures. Since these companies had their own unique policies and procedures for managing projects, quality control, quality assurance, etc., a considerable amount of time had to be expended on retraining experienced appointments who might be reluctant to discard some of this same experience. But, in general, some industrial experience was preferable to none.

Acceptability: personality and work orientation

In addition to academic qualifications and experience, these firms also imposed various definitions of what they considered acceptable with respect to personality, motivation and inter-personal skills. Though there were some variations, three criteria kept recurring in interviews with managers. These were the individual's personal work ethic, their ability to fit in with existing staff and the fit between their personality and the job on offer.

Work ethic

Managers were looking for candidates who displayed an enthusiasm for work, a willingness to become involved and to take on extra work. Personal preferences and experiences, often of a traditional kind, came through in these decisions. For example:

> 'Normally I would go for lads from the country because they are quieter, they are easier to get on with. Lads from the town tend to be, I don't know what it is, they are more outgoing and they'll tell you to "Fuck off" a lot sooner than the lad from the country and they won't be as co-operative . . . Like I'm from the country myself so you would have more in common with them. So when you have more in common with them you get on better with them and they'll do a lot more for you.'
>
> (Andrew, senior technician, Mertel Telecommunications)

Andrew's view was not unusual in that his concern was with the potential cooperation. A supervisor in Trojan Electronics also stressed the importance of cooperation though in this case his 'lads from the town' were 'touring student types':

> 'Once guys are workers and they have the technical qualifications that's the kind of fellow I'd go for. Some student type people might get it into their head "there's no job for me here, I'm going to go abroad for a few years and tour around". Nothing against that . . . but a guy that goes over to England, he works on a building site, he walks into you and shows you that he wants a job and that he worked and that he also got his diploma and whatever. I think he'd be a better kind of guy for the job we have here.'
>
> (Shane, technician supervisor, Trojan Electronics)

Ability to 'fit in'

A second required characteristic was an ability to fit in with existing staff. This decision depended on the manager's understanding of the personalities of his subordinates and a calculated guess as to that of the candidate and how the two would blend. Some of the managers whom I interviewed explained their approach as follows:

> 'When I was doing my share of recruiting, I just have a very simple guideline, if you are going to hire someone what you want is that very happy combination of someone who knows his stuff and happens to be a nice person, simple as that. If you can hire that all the time without specification then you are on a winner. If I put it more complete, you want someone who will fit into an informal working environment, someone with initiative and a sense of personal responsibility, someone who likes a challenge.'
>
> (training manager, Trojan Electronics)

> 'Well I would say that when I would be recruiting . . . working in a company you would know the type of person that would fit into a company. And when you are interviewing you actually tend to look for the people that you would perceive would fit into a particular job within the company.'
>
> (personnel officer, Harford Laboratories)

These decisions produced what other employees saw as particular 'types' that their employer tended to 'go for'. Typically, these were academically talented, had strong inter-personal skills and were able to work alongside others without being unduly arrogant. As Ciaran, a chemist in one of the firms explained: 'They go for technically strong people. They also go for people with good interpersonal skills, quite pleasant and easy to get on with.' (Ciaran, chemist, technical operations, Kenine Pharmaceuticals).

Personality–job fit

There was also a belief that certain personality types were more suited to some technical roles rather than others. This was mostly commonly expressed in the view that certain personalities were more suited to R&D than to manufacturing work. Whether or not this is scientifically verifiable is of less importance than the fact that this belief actually shaped selection decisions. For example, manufacturing jobs required individuals who could handle pressure and communicate clearly with others:

> 'Well within a manufacturing environment you've got to have someone who is . . . its no good having a guy who is suitable for a research lab locked away in a corner. You need someone who can communicate well with people, who is level headed in the sense that he isn't prone to peaks and troughs in temperament because you are dealing with a lot of people, a lot of production people. We're primarily a support function so you've got to have someone who can live in that environment and not end up creating a situation where it's a "them and us".'
>
> (engineering manager, Trojan Electronics)

Along with this, there is the ideal type of candidate who displays either the required flair, a 'technical mentality' or a capacity for a single-minded approach to problem solving. The quality control supervisor in Kenine explained that to get such people he would ask if they were interested in playing with computers, if they had a home computer or anything else which would indicate an interest in 'something that's finicky and tricky and tedious'. A manager in Trojan described the importance of mental stamina and a capacity for logical thought in selecting technicians who would work in PCB repairs:

> 'For instance, technicians troubleshooting printed circuit boards, especially some of the later ones, is a bit like eating kilograms of dry bread. It's a relentless flog through logical connection. You need someone who has a certain

amount of determination and an extremely clear thinking mind not to go round the same loop about five times unwittingly. So you have in mind a sort of specification but it's not written down.'

(training manager, Trojan Electronics)

A surprising feature of the recruitment process in general, and the selection decisions in particular, was the use of traditional methods. Selection decisions were made on the basis of impressions created during two or three-hour long selection interviews and an informal tour of the plant. 'The best' were those chemists, engineers and technicians who made the best impressions on these occasions. Though the process may appear rather old-fashioned these employers insisted they could still distinguish between the good and the bad on the basis of their experience. In this respect the process of recruitment into positions of 'service' is remarkably similar to that of 'waged labour'. Blackburn and Mann (1979) found in their study of manual workers employed by nine firms around Peterborough that the employers believed they could distinguish between 'better' and 'worse' employees. In terms of social criteria the better employees were 'responsible' and 'had discipline'. They could work autonomously and conscientiously without being forced and they had high levels of concentration. Above all they displayed a willingness to cooperate with authority (1979: 107–108). The irony of this particular study was that these characteristics were required for jobs that required less skill than what these employees exercised when driving to work!

The firms studied here used the same criteria even though it was for highly educated technical employees. They too required cooperation with authority. If this seems rather incongruous in such modern hi-tech settings it should be remembered that the problem of managerial control is not one that disappears with the advent of modern production technology (Gallie 1978).

Conclusion

In contrast to their initial experiences in Ireland, the rapid changes in the Irish labour market enabled these hi-tech employers to rely on the sorting mechanisms of the Irish higher education system for the purpose of pre-selecting individuals into different technical occupations. This practice fits with existing research on engineers in the Irish public sector (O'Canainn 1995) and in the electronics industry

(Wickham 1989; Murray and Wickham 1987). The overall pattern indicates that access to technical jobs became rigidly linked to educational qualifications to a point where technical employment was overwhelmingly dominated by qualification-based occupations. That is to say that the occupational divisions were constructed by the national system of higher education system. The craft traditions that existed in Ireland did not shape technical work in the same way that they had in Britain – at least in the British Aircraft Corporation (Smith 1987). The lack of technical employment and, with that, the lack of a clear definition of what constituted a 'technician' (OECD 1964) left it open for the Irish educational system to impose its own definition of what constituted various technical occupations. This pattern of recruitment, which produces an educationally stratified hierarchy of technical labour, is characteristic of what Meiksins and Smith (1993) term the *estate* system. Under this model technical occupations are primarily constructed through educational qualifications, as in France. However, the employment structures for engineers and chemists in the MNC plants also fit Meiksins and Smith's (1993) 'managerial model' of technical employment (Chapter 5).

Despite the propitious labour market conditions, there was no evidence that employers sought to undermine these labour market definitions in shaping occupational boundaries. Those with engineering degrees were not hired into technician roles. Employers were reluctant to do so because of the difficulty of satisfying future demands by these individuals for promotion into their 'rightful' role. While these firms clearly benefited from rising standards within the labour market they were also constrained by the norms established through previous recruitment practices. In this respect, educational qualifications appeared to be a relatively secure labour market asset. However, there was evidence of qualification inflation within individual occupations. Like other processes of inflation, this depreciated the value of certificate and City and Guilds level qualifications. In this respect, educational qualifications were not a secure asset within a dynamic labour market. Savage *et al.* (1992) may be correct when they argue that qualifications are a more useful asset than organizational authority but this does not mean that their value cannot be challenged by employers. Should employers decide that they require more qualified staff than before, or should labour market conditions allow them to select even more qualified candidates, then the previous (job) value attached to educational qualifications will be reduced. In sum, educational qualifications cannot be viewed as a secure, inflation-proof currency within the labour market.

Randall Collins reported in his important study, *The Credential Society* (1979), that educational qualifications are most heavily emphasized within organizations that are governed by a process of normative control or cultural socialization. It is not therefore surprising to find that these hi-tech firms selected technical employees on the basis of certain social criteria in addition to academic qualifications. Managers selected candidates in accordance with criteria that fitted with their conception of acceptability. This led to judgements about the candidate's personality, motivation and interpersonal skills. Such criteria are now accepted as being a common feature of the process of selecting technical staff (Winstanley 1991; Causer and Jones 1993) even though their relevance is not immediately obvious (see also Hanlon 1994 in the case of accountants). Having the 'right attitude' (as distinct from being a 'touring student') was important because it implied that the new employee would accept management's right to exert social control within the organization. As Maguire (1986) has shown, employers can afford to be highly selective where there is an over-supply of labour. In this instance, they could choose those who they perceived to be precisely what they required. As leading hi-tech MNCs, they could attract and select from the cream of each cohort produced by the national educational system. Moreover, the uncertainty involved in recruiting new employees was reduced not only by rising standards in the labour market but also by having an 'inside track' with various universities and colleges. Contacts with academic staff in various universities and colleges enabled them to obtain character references along with information on technical knowledge and competence. The emphasis placed on social criteria, and the choice available to these employers, supports Fox's (1974) argument that the work relations of high-trust high-discretion employees is shaped to a significant extent by the work orientations of these employees. To put it another way, it was clear from the recruitment practices of these employers that they did not have to rely on their ability to monitor and control these employees in order to extract work from them.

4

MARKETS, PROJECTS AND ROUTINES

Introduction

The work of those employed on service contracts tends to have more autonomy and discretion when compared with those employed as waged labour. This is one of the key points of differentiation between these two basic forms of employment relationship (Fox 1974; Goldthorpe 1980; 1995). Goldthorpe (1995) argues that the conditions of employment are more important in defining the service relationship than those of the work situation where there is a discrepancy between the two. This argument is presented in response to the claim that the work situation of professional and managerial employees is not (or no longer) of the kind specified by the service contract (Crompton 1980; Savage *et al.* 1988). What I have set out to do in this study is to examine whether this is indeed the case. Accordingly, the question addressed in this chapter is whether the presence of bureaucratic mechanisms, especially those that lead to greater marketization, has reduced the levels of autonomy and discretion that are believed to be typical of service-class employment.

To recap, the research by Whalley (1986a) and Crawford (1989) found that much of the work of engineers revolved around the handling of projects and other responsibilities. Engineers, according to Whalley (1986a), should be categorized as trusted workers since much of this work drew on personal capabilities to initiate, co-ordinate and take responsibility. Whalley (1986b) also showed that managerial strategies for technical work may vary between market exposure and market insulation. On this point, Whittington (1990; 1991a; 1991b) found that R&D work in the UK was increasingly exposed to product market pressures. Many R&D units were restructured into decentralized profit centres or established as separate divisions in an initiative to force them 'to be commercial'

(1991a: 96). In this changed environment professional staff can no longer enjoy previous levels of job security or discretion. Causer and Jones (1990; 1996) have, as indicated in Chapter 1, modified the service contract perspective with the argument that the high levels of trust and autonomy are constrained by certain controls that management have developed in order to control the technical work process. These include project teams in which managers work alongside technical specialists, a general fusion of managerial and technical roles which ensures an effective means of performance evaluation and a collegial style of management. Since many of these controls refer to the structure of managerial authority, rather than the organization of work *per se*, they will be discussed in the next chapter. Nevertheless, their research raises the question of the kinds of controls employers find appropriate when organizing technical work.

The second reason for examining work is derived from the debate over the role of MNCs in the new international division of labour. According to the NIDL thesis, MNC branch plants only undertake basic manufacturing, or assembly operations, which have limited requirement for skilled work. This is because MNCs locate the more labour intensive and less skilled parts of their production process in less developed countries where labour costs are cheaper (Frobel *et al.* 1980; Henderson 1989). If this argument is correct then it might be expected that these graduate workers would feel over-qualified for branch plant work. Furthermore, they may believe that their knowledge and skills are under-utilized and so may have difficulty in finding work that stimulates their interest.

The first part of this chapter describes the formal organization of technical work. It will commence with an account of the technicians' work before focusing on the nature and significance of projects and project management systems in the work of the chemists and engineers. The second part analyses the experience of technical work in MNC branch plants before closing with an assessment of the implications of management-led changes in the organization of technical work.

The organizational division of technical labour

In order to understand the organizational division of labour it is necessary to make a basic distinction between those technical workers who worked in 'line' functions, such as production, and those in 'staff' functions that provide support services for production. Line functions refer to those roles that are directly involved in the manufacture of the firm's products, that is, in production roles. The distinction between line and staff functions is important in explaining the different levels of autonomy between technicians and engineers or chemists. Those employed on the line in production were part of an interdependent work process. Those in staff positions were much more likely to be involved in projects. Though these might include 'troubleshooting' problems in the production process they generally had a greater degree of autonomy in their work.

The work of technicians

While a major effort was made to match firms from two different industrial sectors, there was a distinct difference between the chemistry and electronics-based firms in the way they deployed technicians (Figure 4.1). The majority of the technicians in the more labour intensive electronics firms worked directly within production where they tested and repaired various products. In the more automated chemistry-based firms the technicians were employed within the supporting quality control function or within the development units. Apart from this, the deployment of technical personnel was broadly comparable in that they either worked in manufacturing support or design and development departments.

In Mertel Telecommunications, the technicians tested and repaired completed telephone sets and exchanges. In Trojan Electronics they tested and repaired printed circuit boards and the completed personal and mainframe computers. Many of these activities involved the use of special test equipment such as oscilloscopes and environmental chambers. Some of these tests, such as the 'burn-in' where components were tested by using a variety of different temperatures and humidity levels to pre-age them, lasted up to seventy-two hours. During this time repeated readings had to be taken and recorded in a manner similar to chemistry experiments in the chemistry-based firms. The operation of this equipment, along with the design and maintenance of the tests, were the responsibility of the quality control departments. An additional group of technicians worked in an area called remanufacturing. Here old models of Trojan personal computers, which had been traded in against newer models, were tested and refitted with fresh components before being resold to the African market.

Chemical/pharmaceutical firms			
Kenine Pharmaceuticals			
Line function	*Technical employees*	*Staff function*	*Technical employees*
		Quality control	Chemists, technicians
		Technical operations	Chemists, technicians
		Planning	Technicians
Harford Laboratories			
Line function	*Technical employees*	*Staff function*	*Technical employees*
		Quality control	Technicians
		R&D	Chemists, chemical-engineers, technicians
Electronics firms			
Mertel Telecommunications			
Line function	*Technical employees*	*Staff function*	*Technical employees*
Production	Technicians	Quality control	Engineers, technical specialists
		Manufacturing engineering	Engineers, technical specialists
		Technology unit	Engineers, technicians, technical specialists
Trojan Electronics			
Line function	*Technical employees*	*Staff function*	*Technical employees*
Production	Technicians	Quality control	Engineers, technicians
		Manufacturing engineering	Engineers

Figure 4.1 Functional location of technical labour

In the chemistry-based firms the technicians were employed in the quality control departments and in the research units. Unlike the electronics firms, the production process in these two firms was highly automated. Both operated sophisticated computer integrated manufacturing processes which had all but eliminated the use of human labour in direct production. Semi-skilled operatives were, however, used to move barrels of chemicals around the plant. Yet the work of the quality control departments, where many of the technicians were employed, was driven by production schedules and was, consequently, quite routine in nature.

Apart from quality control, most of the technical employees worked in staff functions where they had a duty to support the production line in addition to working on projects. This included the manufacturing engineering departments, and the specialist technical departments in Kenine, Harford and in Mertel. While the latter three, the technical operations department in Kenine, the R&D unit in Harford and the technology unit in Mertel, had certain support duties for the production departments their primary role was to develop and introduce new products and processes. Consequently, these departments contained the largest numbers of chemists, chemical engineers and engineers with some technicians and technical specialists (Figure 4.1).

Most of the technicians' work in Kenine Pharmaceuticals, a low-volume high-profit manufacturer, was in the quality control department where they conducted in-process testing on batches of pharmaceuticals. The technicians tested the quality of the various chemical ingredients and reagents ('powders') at different stages of the production process. The chemists in the quality control department had the responsibility for testing the quality of the final products. The nine technicians were divided into three shifts typically with one senior technician on each shift. Testing routines were alternated between those on shifts through an informal arrangement to alleviate the monotony. Two technicians were employed in technical operations where they helped to set up experiments and two more in the planning department where they helped establish and record production schedules and production output.

Quality control work in Harford Laboratories was exclusively technicians' work. For the most part the eight technicians were divided equally between the two manufacturing sites but the most junior technician was occasionally redeployed to the other site depending on which site was busiest. The largest area of technician work in Harford was in the R&D unit where there were about thirty-five technicians. Of the four technical specialists in the firm, three worked in R&D. The main difference between the technicians' grades was that the senior technicians were more experienced and were expected to train new technicians where necessary. Technical specialists were supposed to have a larger role than a senior technician by having some project work. One technical specialist, who worked in R&D, stated that this did not actually occur in practice though some did have additional duties such as membership of the Health and Safety Committee.

Managerial control at work

The work of the engineers and chemists in the R&D or development departments was, as expected from other research (Whalley 1986a; Crawford 1989; Causer and Jones 1996), dominated by projects. In the smaller departments, such as technical operations in Kenine Pharmaceuticals the chemists and chemical engineers had to combine this work with routine duties such as maintenance of the production process and 'troubleshooting'. The same mix of projects and duties constituted the work of engineers in the manufacturing engineering departments in the electronics firms, though they were expected to give a greater priority to any problems that emerged in production. None of the technicians were responsible for projects of this kind. The roles were not therefore entrusted by management with the same degree of responsibility as that of engineers or chemists (McGovern 1996a).

In contrast to the firms described in other research, these MNC branch plants used formalized project management systems to control and coordinate the work of their professional technical staff. These company-specific management tools provided a direct link between technical work and market requirements. In other words, they formalized and made explicit the kind of market exposure strategy described by Whalley (1986b) and Whittington (1990; 1991a; 1991b).

Project management systems

Formalized project management systems were used by all four companies. Unlike the companies described by Whittington (1990; 1991a; 1991b) these MNC branch plants had used this procedure from the time they had established in Ireland. Harford Laboratories was the exception here, possibly because it was established some time before this managerial practice came into fashion (see p. 65). The project management system had two functions. The first was to organize and coordinate the work of those involved in the project. Those involved in the team projects had less autonomy than when they worked on individual projects, tasks or duties because they had to submit to an extra layer of authority and allocate some of their time to the team's work. In theory, the project management system defined the work of employees which would then be monitored and evaluated through the appraisal process and subsequently rewarded through a system of performance-related pay. In practice, the implementation of both the project management system and the appraisal process was constrained by the nature of the work. These constraints will be discussed in detail in the next chapter while the limitations of performance-related pay will be described in Chapter 6.

The second function of this management tool was to bind the work of the technical staff directly to the corporation's product market requirements. In Bailyn's (1985) terms, this meant that the professional staff had little *strategic autonomy*, that is, the freedom to select the projects they worked on. This decision was made through product market research. Projects could only be undertaken after they had been formally assessed and approved by management. Development work was restricted to that which would be commercially viable. For instance, modifications to new or existing products would only be undertaken if demanded by customers. Similarly, there were numerous projects to change the production process because this would help to reduce production costs. The professional staff did, however, have considerable *operational autonomy* so that they could control how the work was to be performed.

Project management systems: two examples of formalization

Depending on the technical sophistication of the branch plant, project management systems were either used for the introduction and adaptation of new products and manufacturing processes in the Irish plants or simply for the transfer of new products and processes from the North American-based R&D headquarters. Though Harford did occasionally develop new products, much of the R&D group's project work, and that of Mertel's technology unit (where they undertook some significant development work), consisted of adaptations to, and introduction of, new products and processes.

The Kenine and Trojan plants did not have R&D groups of similar stature. Consequently, the projects undertaken by their chemists and engineers were limited to the introduction of products and processes that had been developed in the US.

To illustrate the degree of control that could be achieved through these project management systems I shall describe those of Harford and Mertel in detail. Project management was adopted in Harford in the mid-1980s when the company's surge in profitability and market share slowed. Prior to this the company had relied on the inspiration of the R&D staff to produce new adhesives that would almost fortuitously have market potential. This inspiration came from a strong academic research tradition that grew up around the original Director of R&D. But by the 1980s the success of this professional scientific approach had fallen to a point where the company decided to adopt a more explicit market-driven strategy similar to that practised in the Harford research group in the US. The firm would no longer rely on a process, which the personnel manager described as being, similar to monks waiting in their cells for 'Eurekas'. The justification for this market-based approach was expressed by the European Vice President of R&D (*Management* 1985: 6) in terms of customer requirements:

> The key to successful research is the invention and development of materials which are needed. Unused inventions are wasted effort. The more closely research people are linked to the market place, the more their output will be focused on what the customer wants.
>
> (European Vice President, Harford Laboratories)

Projects could originate from two different sources. One was where there was a potential to develop a new application for an existing adhesive possibly by making minor adaptations. Though Harford was a market leader, there was an emphasis on tracing gaps among its own range of products and producing appropriate new ones. The other, and larger, source of projects was where a specific need was identified among its customers. This could arise from interactions between the customers and the Commercial Development section of the New Business Development (NBD) group. At this stage most of the research effort would be put into defining the need and what the characteristics of the new product might be. This phase would culminate in what was termed a DIN or Dublin Identified Need (Figure 4.2). The results would then be presented at a

Harford Laboratories	
DIN	Identification of new product
DAN	Project team established to specify features of the new product Samples delivered to customers
D-Project	New Project team to prepare for product launch
Mertel Telecommunications	
Gate 0	Identification of new product
Gate 1	Project team established to specify features of the new product
Gate 2	Prototypes delivered to customers
Gate 3	Acceptance by Marketing group
Post Gate 3	Production of new product

Figure 4.2 Project management systems in technical work

meeting of the NBD and the European Board of Managers who generally met once per quarter. They would decide the commercial feasibility of the projects and prioritize them accordingly. The project, if considered feasible, would then become a DAN (Dublin Approved Need) and a chemist would be assigned to the project as project leader. Chemists would usually have one or possibly two technicians working with them depending on the weighting given to the project.

During this stage a considerable amount of laboratory time would be invested in finding out whether the project was feasible technically. If it was, samples would be produced and given to the customer who had requested this particular product. The project leader would then engage in discussions with the customer through the marketing department. If the response was positive the project would then be declared a D-Project and a project team would be formed on a matrix management or cross-functional basis. This usually included the project leader, the new product development manager, and representatives from the commercial development and production departments. The project members would assume responsibilities that would not usually fall within their departmental roles and would report to the team leader rather than to their departmental managers. This phase terminated with a product market launch of the new adhesive.

For each phase, project plans would be drawn up with details of the target dates and the steps to be followed to reach these deadlines. Weekly project review meetings were organized where reports were given on the current status of the project, the problems that arose and the actions that were taken. The technicians and chemist also recorded the number of hours they spent on the projects and the costs were built into the project budget. These 'labour costs' were based on estimates of how much the time of these employees cost the Irish plant. This was considerably more than the salaries. This methodology for the handling of projects was also applied to the introduction of new products or equipment into the manufacturing process. In such cases a representative from manufacturing engineering would take the place of commercial development in the project team.

The project management system used in Mertel Telecommunications (Figure 4.2) and indeed in the other firms was quite similar to that of Harford. However in Mertel the project leader had to sign an informal contract that was written up for each 'Gating' phase. Each phase of the project ended in the so-called Gating procedure. This was a meeting to review the status of the project before being approved for the next phase. At Gate 0 a potential new product was identified either by marketing or by the technology group itself. Funds were then authorized to assess the feasibility of the project. At Gate 1 one of the development groups took the responsibility for delivering a defined product whose specifications had already being identified in Gate 0. A schedule was devised and signed by the project leader. This phase consisted of an approved product specification, an approved project plan and an approved test plan. The bulk of the design and development work took place then between Gates 1 and 2. At Gate 2 successful prototypes were produced and offered as samples to potential customers after verified tests. If the response was positive from the field trials the project was gradually transferred to the new product introduction group from the manufacturing department. This group then followed a similar procedure to put the proposed product into full-scale manufacturing. At Gate 3 the marketing department either accepted or rejected the product depending on the results of the customer trials. If successful, the product went into full production after Gate 3. As in Harford, the project was budgeted within the technology unit with estimates of 'labour costs'.

The project management systems were not the only formalized policies relating to the organization of work though they were the most significant. The chemistry-based firms also had to have formalized policies and routines to fit in with the requirements of external bodies such as the Food and Drug Administration Agency. These included standard operating procedures in the area of quality assurance and in production (good manufacturing practice). These policies were rigidly enforced because failure to pass on-site inspections by external bodies would have disastrous consequences. In any case, contaminated medicines or environmental damage from the production process would leave the plant open to legal action. In this way, these policies and procedures added a further layer of bureaucratization to the work of the chemistry-based firms.

The experience of branch plant technical work

Following the claims of the NIDL literature the first issue addressed in this section is that of the compatibility of branch plant work with the expectations that these employees might have on the basis of their qualifications. Specifically, we ask whether technical staff are over-qualified and under-utilized in such firms. Following on this I examine the views and experiences of managers and staff on the levels of repetition in technical work.

Skill utilization and organizational integration

It might be assumed from the NIDL thesis that university graduates would be both over-qualified for their roles and, consequently, under-utilized in their jobs. Two specific items in the questionnaire survey were designed to obtain information on this matter. These asked respondents to give their views on the fit between their work and their qualifications and on the extent to which their technical skills were engaged. The aim was to assess the relationship between the skill and knowledge that resided *in the person* and that which was required by the job content (Clark *et al.* 1988: 90). The aim was to capture information on the subjective experience of work, the feelings and perceptions of employees about the fit between their qualifications and skills and what was actually utilized within the firm. This was considered preferable to one that would attempt to develop an objective measure of the levels of skill in jobs and in different qualifications.

The questionnaire survey revealed that more than three-quarters of the respondents (77.6 per cent) claimed that they were appropriately qualified for their jobs (Table 4.1).

Table 4.1 Over-qualified, appropriately qualified or under-qualified

Company	*Kenine* %	*Harford* %	*Mertel* %	*Trojan* %	*Total* %
Over-qualified	22.2	25.8	13.1	13.5	18.1
Appropriately qualified	77.8	71.0	80.2	81.1	77.6
Under-qualified	0.0	3.2	6.7	5.4	4.3
Total	100.0	100.0	100.0	100.0	100.0
N=	18	31	30	37	116

According to these employees, their jobs were appropriate for people with their qualifications. It may be that in answering this question the respondents were drawing upon the existing labour market conceptions of what an appropriate qualification was for a particular job, as referred to in the last chapter. Here further analysis of this data by level of qualification revealed that a significant minority (30 per cent) of those with postgraduate qualifications stated that they were over-qualified for their current position (a cross-tabulation of respondents' level of qualification by their response to the question on the match between their qualifications and their jobs). These respondents were predominantly chemists based in Harford and to a lesser extent in Kenine (Table 3.1). This explains why one-quarter (25.8 per cent) of the Harford respondents felt they were over-qualified (Table 4.1). In contrast, the analysis of this data by qualification level found that only 12.5 per cent of those with degrees felt they were over-qualified for their jobs. In relation to the use of technical skills and knowledge there was a notable difference between the two industries (Table 4.2). Approximately two-thirds of those in the chemistry-based firms (Kenine 66.7 per cent, Harford 67.8 per cent) stated that the use of their technical skills and knowledge was 'about right' as opposed to slightly more than half of the respondents in the two electronics firms (Mertel 55.2 per cent, Trojan 52.7 per cent). This difference can probably be explained by reference to the actual range of advanced technical activities carried out within these branch plants. Harford Laboratories, as already indicated, contained a genuine *research* and development centre. Here, at least, one of the Ph.D. chemists stated that his job provided him with a suitable challenge:

Table 4.2 Use of technical skills/knowledge

Company	*Kenine* %	*Harford* %	*Mertel* %	*Trojan* %	*Total* %
Far too little or too little	33.3	32.2	41.4	41.7	37.7
About right	66.7	67.8	55.2	52.7	59.7
Too much or far too much	0.0	0.0	3.4	5.6	2.6
Total	100.0	100.0	100.0	100.0	100.0
N=	18	31	29	36	114

Note: There are two missing values

> 'I would have to say so, yes is the straight answer. I'm very happy with it in the sense that we get good opportunities to study and to be creative and there's a carrot there for you when you are innovative and creative and you will be rewarded for it and I think that represents a challenge in its own right.'
>
> (Peter, senior chemist, R&D Harford Laboratories)

This contrasted with the situation in Mertel Telecommunications which was the only electronics firm to have something similar to a design and development function. Even here much of the engineers' work was actually taken up with dealing with mundane day-to-day problems rather than on challenging projects that only occasionally appeared:

> 'But in terms of being actually tested here or in terms of your ability to do things, I think that no we're not over-stressed here. Ideally I'd like to be involved in less of the day to day . . . dealing with departments and keeping a project rolling. That takes up an awful lot of time as opposed to the actual technical challenge of the thing. Technically we're not challenged. Some people are and some people aren't. It depends on the project you get.'
>
> (Barney, engineer, technology, Mertel Telecommunications)

The differences between the two industries also emerged in the interviews. Those in the electronics firms were less satisfied with the nature of their work. Even in the technology unit in Mertel Telecommunications work content was a source of dissatisfaction. Incoming graduates, including those who had worked in the company as placement students, had learned to revise their expectations of the quality of work on offer:

> 'A lot of the development work that really happens is modifications on all American products so you are not making a product from scratch and you are not coming up with a concept. You are just basically taking an existing product and tweaking it to meet your requirements . . . The first time it's a challenge to design a transmission change on a circuit pack. . . If you have to repeat the exercise again in another couple of months on a different pack it is much the same thing and it gets repetitive.'
>
> (Barney, engineer, technology, Mertel Telecommunications)

This problem was one of which managers were aware in Mertel Telecommunications especially as it had led to some engineers leaving for positions in other firms (see Chapter 7). More generally, the routine nature of the work in the electronics plants had implications for the day-to-day motivation of the production technicians and consequently for their supervisors and managers.

The lack of 'quality work'

Existing research on the motivation of scientists and engineers reveals that intrinsic rewards, such as work content, are of critical importance. An organization's ability to provide such job content is in turn closely associated with the perception of it being a good place to work. Jones (1996) reports that the provision of an innovatory climate and job satisfaction are the main determinants of an organization's reputation among R&D staff. Here the two electronics firms did not fare so well. Both the technicians and the engineers complained frequently about the lack of 'quality work'. The provision of such work was the biggest problem that their managers and supervisors had to deal with in order to maintain their morale.

According to the interviews, quality work was characterized by a bundle of inter-related features. First, it was of sufficient technical complexity to challenge individuals who had specialized in electronics during the course of their third-level education. Second, it provided an opportunity to learn new information such as the 'quirks' that came from working on new or different products and processes. Such knowledge represented an investment in human capital that was not only important in itself but also for career advancement purposes. The third feature of 'quality work' was its capacity to capture the attention of the employee by engaging their knowledge and skills in a manner that raised their interest and personal productivity. As a Mertel engineer explained: 'the best way to exploit my potential is to give me the work that I want to work on'.

It was this third aspect that was of particular concern to the managers and supervisors. Unlike those in the chemistry-based firms, they were much more constrained in their capacity to provide such work. Much of this could be attributed to a production process that was based on a variation of the assembly line principle (and some cellular manufacture). In Trojan Electronics much of the technical activity, especially after the redundancies (see p. 79), was at the technician level. In other words, it consisted mostly of repetitive testing and repair activities. For the more experienced technicians, this work had become quite monotonous. The monotony was compounded by the failure of the corporate R&D to produce new products. In the past the arrival of a new product usually led to changes in production methods, new testing procedures, and new types of faults. All of these provided a fresh technical challenge.

The problem was most acute in the board repairs group. It contained the older and more experienced technicians who had learned all that there was to know about the faults and quirks of the existing products. Their supervisor expressed the fear that if the situation continued it would become a major problem as his staff would be 'running out of things to work on which they would see as being a challenge'. By contrast the technicians who worked in the remanufacturing group were younger, less experienced and therefore not so familiar with the company's products. Many were hired within the previous three years from Regional Technical Colleges after Trojan stopped producing its own in-house technicians. Their supervisor was quite aware of the advantage of having technicians who were still learning the practical side of their craft:

> 'So they're new and they're gaining experience all the time and they're very keen to be learning and to keep impressing

> to get on . . . Maybe in a couple of years time it could get very difficult. Once people become stagnant in the area it causes bad feeling, they get pissed off. It's very hard to get technicians, people who have qualified to be doing the same thing every day, day in day out.'
>
> (Shane, production technician supervisor, Trojan Electronics)

The two technician supervisors tried to introduce job rotation to deal with this problem. Not only were they under pressure from the technicians to do this, they were also being pushed by their own managers to give the company some flexibility in the event of illness or turnover. Furthermore, as the business manager explained, the company did not want to end up in a situation whereby all the detailed knowledge of one process or product would be held by one individual. But even here the supervisors had limited room in which to manoeuvre. In the repairs supervisor's view, the only solution for the more experienced technicians was to move them to different work by recommending them for positions in other areas. This then gave him the freedom to take in new technicians, ideally by promoting operators who had obtained the appropriate qualifications. The firm's recent difficulties meant that there were few such opportunities. The supervisor of the repairs group faced an additional constraint. His technicians did not want to move to the repairs group because they would have had to work on 'old technology' that they had worked on some years previously. To even suggest such a move was an insult because they saw it as a step backwards in technology when they wanted to move forwards. Similarly, those who worked on the more complex mainframe computers (VS systems) would not consider a move to PCs:

> 'The PC don't mind moving to the VS because it's a better technical challenge but going the other way is seen as a backward step. That's the problem. I have five tech 4s and five tech 3s. So those guys would see that as going backwards. It's bad enough that they can't go forwards but at least they don't want to be going backwards.'
>
> (Harry, technician supervisor, Trojan Electronics)

The biggest worry for these supervisors was that Trojan Electronics was no longer an exciting technical company. The lack of new products was contributing directly to a drop in morale and, what was worse, the future did not look promising because the corporation had begun to experience a dramatic drop in sales and profits.

A similar problem also existed among graduate engineers in Mertel Telecommunications. Here the problem of providing quality work was compounded in the technology unit by a corporate hiring freeze. The quarterly performance planning meetings had begun to turn into a forum in which engineers asked for moves into other areas. However, the hiring freeze meant that they could not obtain such moves because there were no new graduates to take up the more mundane tasks that they would leave behind. Since these engineers had limited knowledge of the product technology, management found it difficult to redeploy them. To do so would require a considerable investment in time and foregone business before they were 'up to speed' in their new role. The systems design manager, for example, would have liked to expose his engineers to a greater variety of work were it not for this constraint:

> 'Sometimes the conflict with that is when you are under pressure to do things quickly which is generally where we are, there is a conflict between that and using the people you know are expert at something and won't make mistakes. Consequently there is a big tendency to do the job with them because you know that they will do the job right rather than swop people around. That can be a problem.'
>
> (systems design manager, technology, Mertel Telecommunications)

The Director of the unit admitted that there was considerable stagnation among the engineers because of the repetitiveness of their jobs. While he would like to introduce a greater element of job rotation, it was not possible to ignore the market-driven nature of the business. The engineers should therefore make their career plans accordingly:

> 'Engineers, I suppose, come to the realization that we are here and that is the way it is going to be. So they make a career decision on that basis. It's not as though they need it. People come in here to me and say "I don't want to work here any more, I've been on it for three years" and I say "Fine, thanks for coming to me, we'll try to work it

in". But it is not something that is going to happen overnight.'

(director of technology, Mertel Telecommunications)

The result was that many of the engineers spoke of low morale particularly among the graduate engineers as they suffered most from the hiring freeze. Not only had they to revise their expectations about the amount of design and development activity in the unit, they also had to revise their expectations about the length of time it would take before they would be involved in such projects. It was against this background that a number of engineers began to leave the unit for work elsewhere.

This situation contrasted sharply with that of the chemistry-based firms. In Harford Laboratories, a firm with its own R&D unit, the technical staff undertook projects that occasionally led to the registration of new patents. This boosted their status not only within the corporation but also within the professional scientific community. The Director of technical operations in Kenine Pharmaceuticals and his section managers had a policy of rotating projects among staff where possible. On top of their routine work they were, as a former chemist put it, 'ensured an interesting project' because there were always 'twenty projects outside the normal day to day that needed to be done'.

The restructuring of work

At the time of the research there were a variety of changes occurring within the plants which impinged either directly or indirectly on the work of the technical staff. Rather than describing the changes in detail this discussion will focus on the types of changes that occurred and their implications for the regulation of the service contract. The first type was an enlargement of the technical role of the Irish branch plant. In Kenine Pharmaceuticals, the technical services unit had been upgraded so that it was now directly part of the corporate research division rather than simply a technical support unit for the Irish plant. This represented an enlargement of the technical role, with more research activities being devolved from the corporate research group in the US. This unit, under its new title of technical operations, would contribute to research within the corporation for the European market. It would now report directly to the director of corporate research rather than to the plant manager. The change itself resulted from a general restructuring of the overall corporation. The reputation of the Irish technical staff was, according to the training and recruitment manager, also a big influence on the decision. Previous successes in introducing new products and processes had raised the technical standing of the Irish plant. The staff were being sent increasingly on corporate projects which, for example, established new plants in other countries. There was now an expectation among staff that they could become involved in much more advanced development work than previously. While it may not lead to basic research work of the sort carried out by the corporate R&D centre in New Jersey, it would lead 'towards development work on a bigger scale'. One chemist felt that there would be more opportunities in the future for the Irish plant to compete for research projects. He envisaged a competition between the various research groups within the corporate R&D division with the main rival to the Irish unit being the corporate research headquarters.

> 'We could become one of the top groups within the unit if we really use our heads. Cause I have seen them all now and I know their potentials, the other firms. I know we have got the best without bragging about it.'
>
> (Ciaran, chemist, technical operations, Kenine Pharmaceuticals)

Changes in the technology unit in Mertel Telecommunications were initiated at local level. Though the technology unit in the Irish factory was part of the corporate research group, much of its work consisted of installing and maintaining new products and processes along with providing technical support to customers in the European region. It had only begun to design its own products, one of which was a branch exchange system that had a successful market launch. Following on this breakthrough a decision was taken at local level to establish a 'strategic technology group' that would produce innovations to meet gaps in the existing technology. One of the senior engineers who was considered to be a 'key talent' within the plant was given control of this section. Its task was to examine the range of products produced by other companies in order to pin-point shortfalls or gaps in the market. Such assessments would then be used to produce proposals for new products that would be presented to management for an assessment of their commercial viability. This was expected to yield more interesting projects which staff could identify as their particular 'babies', since they would have the opportunity to be involved in the actual conception of new products.

The second type of change reflected the continuing search by management for more efficient ways of working. Here the emphasis was on fashionable managerial philosophies such as Total Quality Management (TQM) and its derivatives. These seek to install 'continuous improvement' within all areas of an organization's business as a core value usually with the introduction of increased employee participation (Hill 1991). Different versions of TQM were adopted by the two electronics firms with the explicit aim of empowering employees to come up with suggestions to improve the quality of their work.

In Mertel Telecommunications the manufacturing department had reorganized the manufacturing process into self-contained cellular work teams. Prior to this, telephone sets and other products were manufactured on an assembly line basis. The assembly line organization of production was divided among a number of autonomous work groups, each with a team leader. The company also introduced 'the coach approach': a strategy to empower the shopfloor workers while changing the role of the supervisor to that of 'group facilitator'. Group facilitators were provided to assist work teams as they organized tasks among themselves. Their role, which replaced that of the supervisor, was to supply resources and advice rather than to monitor and control.

There were additional plans to empower the production technicians by changing their role from one that was dominated by 'hands on' product testing work to one where they would be involved at a more systemic level in quality control. In their new role they would tackle problems in the overall quality process rather than actually carrying them out within the production process as before. Their work would become more of a product support role. This, in turn, had been work that was previously carried out by the manufacturing engineers. There were also proposals to continue the enskilling process by moving the product support engineers in the Manufacturing department to more developmental type work on the production process. The director of manufacturing claimed that this initiative would also help to provide the technicians with more challenging work as well as improve on quality:

> 'On a commercial basis it's more cost effective for me. It frees up a lot of the people to do other things. On a development basis, it develops the actual technicians who

> can be a very demoralized and boring lot of people because they are doing the same thing day in and day out and no one really understands what they are doing anyway in the first place. And they'll tell you that themselves.'
> (director of manufacturing, Mertel Telecommunications)

A more full-blown form of Total Quality Management called the Quality Leadership Programme (QLP) was implemented in Trojan Electronics alongside other corporate-wide changes in an effort to minimize costs and improve product quality. These changes were themselves part of a corporate-wide initiative to bring the company back to profit as it had reported losses in the previous year. The QLP initiative in Trojan also had the basic aim of encouraging all employees to take responsibility for quality issues in their immediate work area. Unlike Mertel, Trojan invested a lot more time in actually training employees in new quality management techniques.

Along with the introduction of the QLP, Trojan Electronics had also restructured its manufacturing activities so that they were now organized around product lines rather than around the work of specific functions. This initiative was aimed at developing a greater market sensitivity within the firm so that it could respond more quickly to customer demands and to changes in the product market. The manufacturing group was divided into personal computer products, vertical storage products and remanufacturing. The result was a matrix system in which each individual product line was supported by the various departments such as engineering, marketing, etc. This reorganization of production also brought the technical staff into closer contact with the pressures of customer delivery deadlines. They now had to support specific product markets as opposed to providing general support to production. This was confirmed by one of the senior managers who spoke of an even greater awareness among engineers about the effects of their work:

> 'I think they're doing the same job but I think there is probably a greater awareness now and a commitment or accountability by the individuals to what their task is. In other words they are still providing engineering support but it's being driven by the market. The business needs are customer oriented and customer focused and therefore the engineer support would be provided in that context, meeting ESDs [estimated start dates], driving to the

Commits to Customers be it quality, be it delivery, be it performance.'

(business manager, Trojan Electronics)

The third form of change encountered in this research, the introduction of redundancies, was viewed as an exercise in 'trimming the fat' rather than as something that related directly to the organization of work. Trojan Electronics, like other major computer manufacturers in the early 1990s e.g., Hewlett-Packard (McGovern and Hope-Hailey 1997), reduced its workforce as computer sales dropped dramatically. A 20 per cent reduction in the workforce was incorporated into a 'delayering exercise' that reduced the number of layers of management. In the Irish plant the reduction was obtained by offering attractive voluntary redundancy terms to 'indirect labour'. Since technicians were directly involved in production they were classified as 'direct labour' and could not therefore apply for redundancy. This was deeply resented by some of the older technicians who had grown tired of the product testing routines and fancied their prospects in the job market. In the end the redundancy package, which was limited to 120 employees, was over-subscribed. The reason was that a large number of engineers decided to leave along with the managers and supervisors with the result that the technical capability of the plant was reduced considerably.

In March 1991, while I was still conducting interviews in the firm, Harford Laboratories announced that they were seeking redundancies in the white-collar and technical categories. This decision, which came from within the Irish plant, was attributed to a drop in European sales over the previous year and in projected sales for the subsequent year. Four voluntary redundancies were sought from among the technical staff.

Conclusion

In recent years an increasing number of organizations have sought to introduce market mechanisms as a means of increasing productivity. According to Whittington and others, this practice has even been extended to professional work (Whalley 1986b; Whittington 1990; Whittington *et al.* 1994). Such initiatives were not encountered in this research because technical work in these MNC branch plants was already market-oriented. The key mechanism for linking technical work to product market requirements was the project management systems. With this practice, engineers and chemists had to become involved in the commercial and financial affairs of their employers (see also Whalley 1986b: 227). Projects had to be financed, budgets had to be monitored and the end product had to have a viable commercial application. It was through such mechanisms that the values and needs of capitalist enterprises were internalized. The traditional tension between the needs of science and those of industrial organizations (e.g., Kornhauser 1962) barely existed in an environment that was aggressively managerialist in outlook.

The evidence presented in this chapter would appear to indicate that the work of engineers and chemists should not be interpreted as evidence of high-trust and high-discretion work roles principally because of the presence of market-driven project management systems driven by profit-related criteria. With this practice, the strategic autonomy of engineers and chemists was clearly limited. As such it indicates that employers did not simply trust employees to serve in their interests. The use of performance appraisals, as the next chapter demonstrates, would also appear to have the same effect. However, as I argue in the next chapter, it would be easy to exaggerate the significance of these practices.

The evidence on the fit between the qualifications and jobs and on skill utilization revealed that the electronics technicians, in particular, were not able to fully utilize their skills and knowledge. This is not an uncommon finding for technicians employed in hi-tech industry (see also Hodson and Hagan 1988). In this case, it was not what might be expected from firms that were selected because they represented the technical elite within Irish industry. What this suggests is that there is indeed support for the NIDL thesis in the Irish case when these elite 'hi-tech' employers were unable to provide work that met the needs of their technical workers. What tends to be ignored by this literature are the management problems caused by this kind of work. The managers of the two electronics firms in particular were constantly under pressure to accommodate demands for 'quality work'. Market pressures, customer deadlines and the time required to train an engineer in a new area all militated against demands for quality work. This generalization does not, however, apply to the chemistry-based firms, at least to the same degree. It may be that the NIDL thesis best fits high volume mass production-type industries of the kind examined by its original proponents such as textiles (Frobel *et al.* 1980) or electronics Henderson 1989).

This finding has two implications. First, it is not accurate to claim, as Goldthorpe (1982) has, that these employees will perform in their employers' interests because of an open-ended moral commitment to the organization. Rather this chapter shows that such commitment may not be so readily available at the day-to-day level when the work is not of the kind expected. That is to say that the nature of the work and the organizational context cannot be completely dismissed from attempts to categorize the class location of professional (and managerial) employees. Second, this implies that a note of reservation must be placed on the employee commitment strategy that is so popular in the human resource management literature (Beer *et al.* 1985; Walton 1985). Much of this literature suggests that human resource policies can lead to a high commitment model of employment. The possible impact of work organization and the experience of work on employee commitment is ignored not only by those who promote such models but also by those who have sought to examine their practice (e.g., Wood and Albanese 1995). This is all the more remarkable given the large number of well-known studies in industrial sociology that demonstrate that the experience of work can have a distinct influence on social integration within firms (e.g., Blauner 1964; Gallie 1978; Lincoln and Kalleberg 1990). The evidence presented here shows that employees who have a high degree of (professional) commitment may actually have this frustrated by routine work. If high levels of employee commitment are to be achieved through various HRM policies then it must be assumed that employees will see fit to numb themselves from the experience of work that is of a monotonous nature!

Finally, of the various changes that these firms implemented two are of significance. First, it was clear from the TQM-type initiatives in the two electronics firms that management sought to exhort employees to contribute freely to decisions concerning their work while simultaneously reducing formal managerial control by removing supervisors. This attempt to empower employees has clear parallels with the high-trust approach described by Fox (1974), notably the efforts to increase employee discretion and to draw on their moral commitment. What this also implies is that management may not always choose to circumscribe the amount of discretion and responsibility they give to their staff. They may actually seek to expand it in their search for efficiency, albeit within certain limits such as those set out in the project management systems described earlier.

Second, the assault on the so-called 'corporate fat' in two of these corporations is contrary to the traditional assumption that such employees have greater job security than those in waged labour positions. Job cuts were aimed specifically at engineers and supervisors in Trojan Electronics and at technicians and clerical employees in Harford Laboratories. This is not something that fits comfortably with the idea of service-class employees having relatively secure positions. It would appear that a concern to protect the job security of the so-called trusted workers was not among management's priorities in their search for profitability.

5

MANAGERIAL AUTHORITY AND BUREAUCRATIC CONTROL

Introduction

If the employment relationship of professional and managerial workers is governed by a relationship of service then we might expect to see this reflected in the form which managerial authority takes within the work situation. Both the structure of managerial authority and the pattern of relations between managers and employees should reflect a relationship that is founded on the moral involvement of the employees. Goldthorpe (1982; 1995), however, has little to say about the nature of authority other than employers simply trust their employees to act in their interests. Fox (1974), however, stated that the management of high-trust high-discretion work roles was conditioned by the work orientation of the role occupants. Since this orientation was one of moral commitment to the organization, detailed supervision was considered to be inappropriate because, as Fox claims, the emphasis was on joint problem solving rather than on the unilateral imposition of procedures. Work coordination was on the basis of mutual accommodation, or adjustment, with open lines of communication and a free-flowing exchange of suggestions, advice and criticism. With regard to those in the 'middle range of discretion', such as technicians, Fox claimed their work situation was subject to a slightly different pattern of control. Though their superiors might insist that their interests were the same, they were nevertheless subject to formal controls and monitoring devices. These served to place technicians, clerical workers and others in a clear position of subordination while simultaneously emphasizing their social distance from their superiors (1974: 30–38).

This chapter examines the validity of Fox's account in the context of recent developments in managerial practice. Two areas are investigated in detail. The first is the possibility of differences in the structure of managerial authority between those in high-discretion high-trust work roles (in this case engineers or chemists) and those at the intermediate level (technicians). In this context it is important to ask whether management are still dependent on these orientations in the context of certain HRM practices. If, as some suggest, the employment relationship of those in the service class is characterized by formalized management controls then it remains to be seen if these have indeed altered the underlying social relations of authority. The second, and related, issue concerns the style of management. According to Fox, managers try to accommodate the orientations of staff. But is this the case? If so, how?

Managerial structures in technical work

Edwards' (1979) distinction between *direct* and *bureaucratic* forms of control is a useful starting point for the analysis of managerial authority. According to Edwards, direct, or simple, control is embodied in the role of a supervisor. Supervisors have the power to hire and fire, to evaluate, promote and discipline as they see fit. Above all they have the power to direct employees as to what tasks they should and should not carry out and in what manner. Bureaucratic control, by contrast, refers to situations where managerial authority is institutionalized in the form of company policy or rule by law. The functions of management are routinized into a set of goals, rules and procedures in such a way that the need for direct intervention by supervisors, managers or capitalists is removed. Control is expressed through the structured design and management of job systems. This strategy originated in attempts by progressive US firms, such as IBM and Polaroid, to apply these methods to non-union white-collar workers though its success led other firms, such as AT&T, US Steel and GE, to apply it to production workers (Edwards 1979 130–132).

Both modes of control were present in the MNC branch plants. Each implemented formalized policies in the areas of appraisal, remuneration, project management, etc. These policies were designed at the corporate headquarters in North America and were implemented uniformly in various plants around the world. This is not surprising since Edwards (1979) has argued that bureaucratic control first emerged in large blue chip corporations of this kind. The use of performance appraisals and project management systems was, however, reinforced to varying degrees by traditional supervisory practices. Here the picture becomes slightly more complex. The variation in the application of direct control depended on two factors: employment status and trade union organization (Figure 5.1).

The production technicians in Mertel Telecommunications and the technicians in Kenine Pharmaceuticals and Harford Laboratories were under direct supervision. Though Trojan Electronics was a non-union company, and therefore free to implement individualized performance systems, it also assigned supervisors to direct the work of the technicians. Typically, the technician supervisor spent more time monitoring work than allocating or coordinating it. While the presence of a supervisor suggested that the technicians were comparable to shopfloor workers, much of their work consisted of set testing procedures that did not have to be allocated on a daily or even weekly basis. All were highly experienced in the use of certain test equipment and so did not require direction from supervisors. Some of the chemistry technicians had even specialized in the use of certain instrumentation to the point where they had become much more skilled than the chemists who were supposedly their technical superiors. In the chemistry-based plants the supervisors' input was most notable when decisions had to be made about the status of various samples. They had to help decide which samples were to be given the highest priority and tested first, and also what the final decision would be in the case of a 'borderline' sample. Similarly, the supervisors in the electronics firms were called upon for decisions on difficult board repairs or other problems that were out of the ordinary flow of work.

The performance review systems that covered the professional and managerial staff were corporate-wide policies. Typically, the professional staff sat down with their managers and formally agreed on a set of work objectives for the forthcoming review period. This process took place annually in Harford and Kenine and approximately every quarter in Trojan and Mertel. Managers in each company were, according to the various appraisal handbooks, expected to agree with each employee on the priority of the different objectives, the target dates for the completion of these objectives and on the criteria for evaluating success or failure. These objectives were then drafted and signed by managers and their employees. The employees had thus given their word to meet the required targets and signed their name to that effect. The following is an excerpt from the work objectives of a manufacturing (test support) engineer in Trojan Electronics:

Industry	**Pharmaceuticals/chemicals**			**Electronics**
Firms	**Kenine**	**Harford**	**Mertel**	**Trojan**
Bureaucratic (Appraisals/projects)	Salaried, chemists,	Salaried, chemists, chemical engineers	Salaried, engineers, chemical engineers	All employees technicians
Direct (Supervisors)	Hourly, technicians (union members)	Hourly, technicians (union members)	Hourly, production technicians	Hourly, production technicians

Figure 5.1 Forms of managerial control
Note: The term 'salaried' refers to the managerial staff in these organizations

> To perform a pivotal role within VS test engineering and to ensure consistent test strategies for all VS; To provide technical support, expertise for less experienced product test engineers on VS hardware design, micro code and operating systems; Responsibility for monitoring European certification requirements for VS products and adhere to the same by R&D, any issues that may happen be given back to R&D; Set direction for improvement of automation of test software for PCL final assembly.

With continuing goals like these, specific deadlines could not be used as a controlling device. Instead progress would be checked at such times as 'end of quarter'. The professional staff also had considerable operational autonomy in deciding how to complete these objectives. Unlike the technicians who only had the autonomy to introduce slight variations into their regular routines, the professional staff had a much wider span of discretion. Their tasks were of a much more open-ended nature and their work was not dominated by routines in the same way as the technicians. This point was reinforced during interviews with managers who explained that it could not really work in any other way. Rather than try to specify tasks and activities in advance, they preferred to map out objectives within broad parameters.

Beyond bureaucratic control

While the use of project management systems and performance appraisals would appear to have circumscribed the distinctive degree of autonomy associated with professional employees, it would be misleading to view the existence of such policies as evidence of effective management control. Indeed in some instances they were interpreted as "bureaucratic" exercises of the "red tape" variety. A technician in Mertel had this to say about the company's 'Managing For Achievement' (MFA) appraisal process:

> 'My understanding of that is MFA stands for "Means Fuck All". Do you want a serious answer to this? My attitude is that it is a load of bullshit. It's a bureaucratic paper exercise. . . It's a procedure, he has to bring me in, he has to go through it.'
>
> (Eddie, technical specialist, Mertel Telecommunications)

More generally, the implementation of these management practices was constrained by the indeterminacy of technical work and by variations in the practices of individual managers. First, the nature of the work was such that the appraisal system could never be implemented with precision. Professional work is frequently of an indeterminate nature (Child 1981). Technical work, including that of technicians, is no different (Barley and Orr 1997). As indicated earlier, this point was recognized by managers when they agreed objectives with staff. Second, work priorities changed so rapidly that it was not possible for the formalized appraisal or project management systems to keep track of them in practice. In the technical support functions, assigned tasks and projects were frequently interrupted by the arrival of problems from the production units, from company customers, or even from other branch plants. Since these were primarily production plants, production problems were always given the highest priority because any interruption in the process could cost thousands of pounds within a few hours. Consequently, project schedules that had been arranged, 'costed' and 'signed' had to be set aside as attention was switched to the new and more immediate problem. Not all of these originated in production. For example:

> 'They fill in your job description, your manager fills in the same thing for everyone "To design and develop products from the conceptual stage to the product stage meeting budgets on target and on schedule". They write that in for everyone whereas at the end of the day nine out of ten of those people that they write it in for mightn't be developing anything, they might be supporting a product out in Germany or out in France. People are moved around a lot depending on what is seen as a hot point at the moment.'
>
> (Barney, engineer, Mertel Telecommunications)

Barney found this to be a frustrating experience. He was particularly interested in design activities and occasionally got the opportunity to work on a project that allowed him to develop his ideas. However, these opportunities were constantly curtailed by instructions to switch his attention to the latest 'hot point'. He left the organization some months later.

In this context it had become common practice among managers to ignore the paperwork and concentrate their efforts on more important tasks. A technician supervisor in Trojan described his situation in these blunt terms:

> 'We hired people six months ago and the job I'd hired them to do then, if I'd given them goals, listed out ten goals for them, I couldn't review them now on those goals because their goals have changed so dramatically it's unbelievable. Right now in Trojan the business is changing so often and people are being transferred around so individual goals are stupid.'
>
> (Shane, production technician supervisor, Trojan Electronics)

Some managers tried to use the next review period as an opportunity to bring the appraisal up to date before allocating merit ratings. This led to a considerable amount of individual discussion and bargaining in which employees defended their failure to meet their formally assigned goals. Attempts to re-write the objectives and goals on a continuous basis were not considered a good use of time. A senior manager in Mertel explained that it would lead to over-planning, something that was just not practical for the work involved:

> 'But it (MFA) is hampered by the fact that the business is changing quite a lot. I've a guy who came back in here after Christmas and he was trying to do some work and all of a sudden he got a frantic call from a customer in Germany. He's out there on site now. So how do you accommodate stop-start type things? Well re-writing it is too cumbersome, I don't agree with that at all. You can over-plan as well.'
>
> (director of technology, Mertel Telecommunications)

This behaviour led to differences of opinion and occasional rows with members of the personnel departments. They insisted on receiving the completed appraisals for audits and for employee records. Despite their exhortations, they found it difficult to convince their line managers that they were not wasting their time. The managers usually responded with the view that they had more important issues on which to spend their time.

The managerial side of this problem is not unique to technical work. Research on managers' practices in a cross-section of major

British-based companies found that the appraisal practices of managers tend to be uneven in quality and inconsistent in implementation. Much of this was attributed to organizational constraints such as the lack of training and support, the failure to evaluate managers themselves on this activity, along with the pressure to meet short-term financial or sales targets. In this environment the priority attached to HRM activities tends to become submerged by more pressing requirements (McGovern *et al.* 1997). Given these problems it would be wrong to assume that the presence of appraisals, and indeed other bureaucratic monitoring devices, has effectively curtailed the high-trust element of professional employment.

The conclusion to be drawn from this is straightforward: we must look beyond the formal policies and mechanisms if we are to understand how technical work is extracted. I make no claim for originality with this argument as it replicates the emphasis that was placed on the informal organization by industrial sociologists in the 1940s and 1950s (e.g., Roy 1952). Yet it is important to emphasize the point when others overestimate the significance of performance appraisals in regulating the work of white-collar employees (Savage *et al.* 1988; Halford and Savage 1995). To that end I shall now describe three inter-related elements of the structure and process of management control in technical work. These include the fusion of technical and managerial roles, the technical background of the managers and supervisors and, finally, the collegial style of management that follows from these structural features.

The fusion of technical and managerial roles

A distinctive feature of the structure of managerial authority was the way it combined both technical and managerial elements for those in managerial *and* senior technical posts. At the senior end of the technical hierarchy there was a *fusion* of the technical and managerial roles (see also Causer and Jones 1996) with the result that many of those who held the position of chemist, or engineer, had significant managerial responsibilities. The supervision of technicians, for example, tended to be a feature of the work of the more senior engineers and chemists. This traditionally management function was also built into the role of project leader. Typically, it was a senior engineer or chemist who was appointed project leader. Only in exceptional circumstances did technicians manage projects. When the project was under way, the other members of the team reported to the project leader in addition to their departmental manager. In this way, each project introduced a matrix reporting structure that lasted until the task was completed. On the larger projects the role of the project manager resembled that of a full-time manager with large amounts of administration, scheduling activities, financial costings, etc., as well as numerous team and customer meetings. The project leader attended meetings with managers from across the organization and, if necessary, with those from sister branch plants. Decisions on the allocation of work and resources, administration, scheduling and negotiations with suppliers and customers were part and parcel of the role of project team leader. Such work was frequently delegated within the team in the case of large projects.

The opportunity to manage a project was eagerly sought by those who wished to be considered for promotion into the manager ranks. Although it might mean additional work the experience of managing a project acted as a stepping stone to a full-time managerial position and ultimately to a career in management. This was all the more important since these branch plants were, with the single exception of Harford, unable to provide extended technical career paths (Chapter 7). Senior technical staff had to either accept that their careers had peaked by their mid-30s – after one promotion – or else look for further promotions by stepping onto the managerial ladder. It was not therefore surprising to find that these employees identified closely with the interests of management. A chemical engineer in Harford used a sporting metaphor to illustrate the strength of this allegiance: 'I would certainly see us as being very much in the managerial half of the pitch, if you know what I mean, probably just outside the penalty box'. This account fits the 'managerial model' of technical employment (Meiksins and Smith 1993) in which technical work is closely aligned with the interests of management as in the US. Though technical employees were recruited into the MNCs in a manner which typified the 'estate model', the fusion of technical and managerial roles in the work of the engineers and chemists and the career structures – which almost inevitably led into management positions (Chapter 7) – ensured that the engineers and chemists identified with management. The result is a hybrid mix of these two models for the professional staff (McGovern 1996a). However, the estate model, which is based on processes of educational stratification, explains the employment of technicians more accurately because their job structures and career opportunities were such that they were unlikely to hold managerial responsibility. For the technicians in Kenine and Harford, their membership of trade unions also indicated that they did not identify so closely with managerial interests.

In sum, the managerial responsibilities associated with projects in these firms tended to be quite fluid as various project teams were reconfigured according to the expertise required. This pattern is remarkably similar to that found in electronics firms located in central southern England by Causer and Jones (1990; 1996). They claim that this practice is not surprising since it fits the 'organic' pattern of work organization which Burns and Stalker (1961) identified in their classic study of the electronics industry. Their description of authority under this organic pattern could easily have been based on team projects within a matrix management structure some thirty years later: 'knowledge about the technical or commercial nature of the here and now task may be located anywhere in the network; this location becoming the *ad hoc* centre of control authority and communication' (1961: 121). In short, the boundaries between the professional and managerial roles were quite blurred under this form of work organization. The delegation of authority to project team leaders had broken down traditional boundaries between managers and their subordinates. As Jones *et al.* (1993: 44) stated in relation to engineers: 'The dividing line between "engineer" and "manager" is imprecise, and it becomes more so as careers progress'. Whatever the trends towards divergence between the class position and life chances of self-employed professionals and managers (Savage *et al.* 1992; Goldthorpe 1995) the conclusion from this research must be that the work situation of *organizational* professionals and managers is quite similar in that they both exercise authority (or to use Marxist language fulfil the functions of capital). The real possibility of moving into full-time management positions reinforces the point.

Technical managers and technical knowledge

The technical managers and supervisors in these four branch plants had, without exception, lengthy experience in technical roles. They had started their careers in technical positions after taking university degrees in science or engineering (certificates or diplomas for the supervisors). Most had only ever worked for one employer. All that they had achieved in their careers they owed to Mertel or Harford, etc. They were classic examples of the 'company man' (I only encountered two female technical managers between the four firms).

All of them believed that their technical experience was essential in enabling them to monitor and regulate the work of their staff. They were all familiar with the standards required by the company and the challenges presented by technical work. This, in particular, allowed them to dictate the norms and expectations with respect to workload and performance. A manager in Trojan Electronics explained that it was precisely because of this ability that they had a deliberate policy of selecting supervisors from the ranks of the technicians:

> 'It's very, very grey in there and for that reason I suppose we have stuck to having the supervisor over the technicians. They're ex-technicians, or they have a technical background themselves, so as to understand the problems that technicians come up with and to eliminate the bullshit factor of a technician when it comes to getting work out of them or getting the job done.'
>
> (personnel area representative, Trojan Electronics)

Furthermore, managers and supervisors could draw on their accumulated knowledge to obtain resources, make suggestions and provide support for those struggling with technical problems. Those who were capable of helping their staff to overcome these problems were held in high esteem. Good *technical* managers were those who 'knew their stuff'. Their ability to suggest new ideas and solutions meant that they could direct their department from a basis of technical authority (see also Clark *et al.* 1988: 133–159). Other managers who had less technical knowledge could still draw on their contacts to solve problems. For example:

> 'It tends to happen naturally that if someone has a problem which they are having difficulty resolving that they would either bounce it off other people within the same group or senior people in the group or even off myself. I wouldn't have the same level of product knowledge as the engineers working for me but having the experience I have I can often suggest a direction to take even as far as "Well let's contact R&D" if the problem is of the scale that I need to be involved from a liaison point of view.'
>
> (engineering manager, Trojan Electronics)

This practice is of significance because it shows that even managers with technical experience may not understand significant elements of the work of their staff. This is important because it adds further support to one of the central arguments of this chapter, namely, that management control is far from complete in technical work. Managers admitted in interviews that they did not know, or in some cases understand, the precise details of the tasks undertaken by members of their department. This was due to the rapid rate of technical change within the industry and the frequent introduction of new products and technologies into their factories – after they had been developed at the corporate R&D headquarters. Their ability to stay up to date with developments in technology was also curtailed by virtue of the fact that they were full-time managers who had little time to read up on these changes. Despite the best efforts of some individuals, their work was almost totally managerial in content. It took a major effort to 'keep their hand in' on the technical side. Even those who were held in high esteem because of their technical prowess found that their knowledge gradually declined. As the years went by, they confessed that whatever knowledge they had gradually became outdated.

In sum, even with the fusion of technical and managerial roles and despite the technical background of the managers they could still only achieve a measure of control. The setting of standards, for instance, was not a straightforward process. This was because the requirement for creativity in technical work was too great while the relationship between activities and outcome was much too unpredictable. The technology was also too complicated and changed too frequently to make it possible for management to aim for detailed control. It is for this reason that Breen (1997: 479) claims that the service contract is based on an informational asymmetry problem since the employer is unable to monitor and so obtain information on what exactly the employee was doing.

Management style: collegialism and professional standards

In this context managers found it appropriate to use a collegial style of management. This is defined by Causer and Jones (1996: 116–118) as a pattern of social exchange in which employees are treated as colleagues, rather than as subordinates, and managers use their technical enthusiasm and competence to stress the common identity that they share with their staff. This pattern of exchange was quite evident in the branch plants. Social interaction between managers and staff consisted of information exchanges, project updates, the exchange of advice and discussions on new technology. What was remarkable was the absence of unilateral commands or orders. Overt displays of authority were not required since both parties were working towards mutual interests. This point was made repeatedly in interviews with engineers and chemists. One engineer, for example, described this pattern as follows:

> 'It's up to you to go and get it done. You will be talking to your manager, like you will meet him in the corridor and you might sit down once per week or so but it doesn't tend to be so structured. It's kind of more informal than formal. And then before every major deadline you are going to have to sit down and work out the various points and see what's done and what's not done.'
>
> (Barney, engineer, technology, Mertel Telecommunications)

It would be wrong to assume that management had to manage through a collegial style purely because of the information asymmetry problem. It also suited their interests as it enabled them to accommodate, and capitalize on, the professional (or bureaucratic) orientations of their staff. That is to say, it was appropriate for employees for whom scientific–technical work was a central life interest. They had already devoted four years of their lives to the study of the subject before subsequently embarking on careers in the area. Consequently, their involvement in their work had definite moral elements (Goldthorpe *et al.* 1968). In short, science and technology were central to their 'life fate'. Nor should it be forgotten that when compared with shopfloor employees, technical workers were not as readily available in the local labour market. As the personnel manager in Mertel Telecommunications explained 'engineers and these people are scarce objects and they are treated differently and nobody would have any objection to that'.

However, the most interesting aspect of this style was the way that managers *expected* technical staff, and technical employees expected each other, to be committed to their work. This process of management by expectation was essentially a form of ideological control. Managers made it clear that as professional employees they were expected to perform to the highest standards and to be devoted to their careers. This in turn was shared by the engineers and chemists themselves. A senior engineer in Trojan emphasized the importance of an individual's professional integrity. This, in his view, was what separated the engineers from shopfloor workers:

> 'I suppose it comes back to engineers or any professional body, the one thing that sustains them more than anything has to be a kind of personal pride or personal integrity, their professionalism more than anything. Because if engineers or professionals have to be chased in the same manner as hourly paid assemblers you have to pose a serious question about their professionalism.'
> (Tony, senior manufacturing engineer, Trojan Electronics)

According to the engineering manager in the same organization, the engineers themselves knew what was expected of them:

> 'A lot of it's due to the fact that engineers themselves understand that they own a certain responsibility or responsibilities, that very much the resolution of any difficulties around those responsibilities lies with them. A lot of people themselves are reasonably well self-motivated and get on with the job.'
> (engineering manager, Trojan Electronics)

Another engineer reinforced this by explaining just how easy it could be to go through the day without doing much work were it not for the expectations that surrounded his performance:

> 'If I really wanted to, really and truly, I could come in here some days and sit around and do very little. Nobody's going to come and look. I'm expected to do what I'm doing and that is more or less the way it is for most people. Probably except for people directly on the lines who are being measured. But anybody outside of that, they're expected to do their job.'
> (George, senior engineer, quality control, Trojan Electronics)

A senior manager in Harford Laboratories described the overall approach to managing these employees as follows:

> 'For most of the people it's a lot of professional pride and we try and recruit people who have that professional pride. We try and give them the environment where they can exercise it and where they have the facilities and the back-up around to do it and we try and personally say "Well done" to those who have done well.'
>
> (Vice President of R&D, Harford Laboratories)

This last account also shows that these employers were not completely dependent on the degree of professional integrity shown by various individuals. They could control their degree of dependence by carefully recruiting those who displayed a sense of 'professional pride'. However, this too was far from being a precise process.

As I indicated earlier, the expectations of colleagues were also important. What was most valued in this collegial environment was the respect of one's peers since they were the severest critics (as in most forms of professional employment). A reputation as a competent and trustworthy engineer required a lot of smart work. Once this was achieved, the pressure was still there to live up to colleagues' expectations. To have a reputation as a good engineer or technician or, ideally, to be known as a 'guru' was a major achievement and huge source of professional pride. That such a reputation should come from your colleagues was, as one engineer explained, 'worth a lot':

> 'I think one of the main things around here is to be seen to be doing a good job by your fellow engineers. I think the best compliment you can get around here would be recognition on the lines of "Oh yeah he's a good guy, he knows his stuff" by your fellow engineers. Your fellow engineers would know whether you are a good engineer or not and that credibility is worth a lot. Your identity is wrapped up a lot in your work.'
>
> (Padraig, engineer, technology, Mertel Telecommunications)

The best way of obtaining a visible reputation was by being involved in high-profile projects that would have a significant bearing on the plant's future. These could make or break the technical credibility of an engineer or chemist because their success or failure could be traced directly back to that person:

> 'I would say to some extent that it is probably the amount of responsibility that is given to each individual. Usually if there is a success story you can pin-point exactly who it was because they have been high profile on that job. You just always know what everybody is doing so that when something is a success you just associate it directly with an individual. It has got to do with reputation with your peers.'
> (Ronan, engineer, technology, Mertel Telecommunications)

> 'It's a purely self-imposed type of pressure or from your peer group. You are working on a project. You take pride to see it through. If you fail to deliver it, and someone else delivers it, your credibility dropped. Not only by yourself and in the group but also by management.'
> (Frank, chemist, technical operations, Kenine Pharmaceuticals)

Though the technicians appeared to be motivated by similar reasons, their levels of productivity were driven by more direct pressures. Production-based tasks differed from those in technical support roles in that there was a greater degree of interdependence in the way their work was organized. A stoppage or problem in one area of production could bring the whole production process to a halt. This contrasts with a situation of low interdependency in staff functions. If a test engineer was having difficulty devising a new operating procedure for Printed Circuit Boards (PCBs) he was not delaying any other employee, especially if there was an existing procedure. Production roles therefore contained more in-built performance pressures. This was especially true for the production technicians in the electronics firms and indeed to those who provided technical support for manufacturing in all four firms:

> 'It's just production, it's different to any other department. In production you have to do the work. You have no choice . . . If we come in at twenty-five to nine, we're put down late. That applies to work as well. They [engineers] could sit down there and chat for the day but we're out there, we're exposed.'
> (Andrew, production technician, Mertel Telecommunications)

Production was therefore seen to be a 'tougher environment' with more rigid scheduling and less choice for employees as to how they could organize their day's work. The production technicians' work came at the end of the assembly process where they tested the operation of the various products and made repairs. Here they had the additional burden of reducing the amount of time it took to test products so that they could be released more quickly to customers. For this reason their work situation was marked by lower levels of autonomy that than of the engineers who had greater freedom to decide what they worked on, as well as when and at what pace. The occupational division of labour within these organizations was such that these decisions were already made for the technicians.

It should be emphasized that this does not suggest that the technicians had weaker professional values than the engineers or chemists. If they had, it could not be confirmed by this research as it was not addressed in the employee survey. However, managers in Trojan Electronics believed that there was a tendency for some technicians to use the complexity of their work as a shield behind which they occasionally 'coasted' (or 'took it easy'). To recall an earlier quote, it was precisely because of the technicians' 'bullshit factor' that Trojan insisted on having ex-techs as supervisers for the technicians. However, when they attempted to change some work procedures they came into conflict with the technicians' sense of standards and professional competence.

'The numbers game'

While interviewing in Trojan Electronics, I discovered that a dispute had developed between the management and the test technicians. This was essentially a conflict between the technicians' standards of work and management's conception of efficiency. It had arisen from the introduction of a series of measures to tighten control over inventories of printed circuit boards and other computer components. These measures were of a classically bureaucratic kind. Areas of decision-making that had been guided mostly by technicians' judgement would now become subject to formalized management policies. What had been governed by tradition would now be governed by 'rational' management procedures.

The PCB repair technicians were ordered to reduce the backlog of 'failed' printed circuit boards that they were supposed to repair. Previously, the backlog of failures were kept in a store room.

The technicians' work consisted of testing and repairing boards taken from this room. The new directive from management was that the technicians should quantify the amount of time spent on a board. According to the production manager, there was enough information available to enable the technicians to decide when it was no longer worthwhile to work on a particular board or problem. Therefore, the technicians should in future calculate when it would no longer be cost effective to fix a board and move onto another.

However, this decision to scrap boards on the basis of a number of 'costed' hours ran against the work ethic of the technicians. They believed that the length of time it took to fix failures was not quantifiable as there were a variety of failure types, including those that changed during testing. To insist on such 'artificial' figures for the cost of a repair would turn their work into a silly 'numbers game'. Their argument was that the extra time spent finding a particular fault, especially if it was a new type of fault, would subsequently save the company huge sums of money. Furthermore they did not want senior managers, who they felt had little appreciation of technicians' work, to tell them how they should repair PCBs. From their perspective, the drive to cut inventories simply made their supervisor look good if he could get the numbers through irrespective of what the faults were. What they most resented was that the changes implied that they were 'no longer being treated as a technical area': the emphasis would be on cost-cutting and output rather than on problem solving.

This difficulty was acknowledged by their supervisor. However, he also felt that some 'lazyish' technicians who had lost interest were hiding behind the arguments of their colleagues. Even so, he had difficulty in persuading the technicians to change their ways. He had held a number of meetings with them but to little avail. Most of them were simply not going to ignore what they saw as their duty as technicians. As one of the more senior technicians explained, they had a certain sense of responsibility to do things right:

> 'Well the way I look at it is, I'm there to fix problems not to fix numbers, I have to be responsible. . . it is my responsibility to do things right and to get the products right and to solve the problems and not just to play the numbers and play the system.'
>
> (Rory, tech 4, Trojan Electronics)

In the end, the technicians were forced to play 'the numbers game' despite their misgivings about its value. They had to accept that the Irish plant's reputation for technical excellence and their own principles had to give way to the cost-cutting requirements of a struggling corporation.

This dispute over the 'numbers game' illustrates one of the key themes of this study, namely, the bureaucratization of work and the market pressures that drive such a process. In crude terms it was a clash between accountancy and engineering. Management wanted the technicians to base their decisions on financial costs rather than on technical knowledge. The technicians wanted management to appreciate their technical knowledge and to acknowledge the possibility that they had the potential to reduce cost over the long term. Here the dispute reveals much about the strength of the technicians' values precisely because they brought them into conflict with management. Though their work may have had lower status than that of engineers and though they may have been subject to more direct forms of control, their values led them to believe in certain standards of work. This supports the notion that technicians should be viewed as an intermediate group. While they considered themselves to be highly skilled they were not treated as a 'technical area'. Fox's (1974) description of this situation is apposite:

> Although in many cases subjected to ideological pressure designed to promote perception of shared goals with their superiors, the structure of their work situation often appears to reveal a conviction on the part of higher authority that they cannot wholly be trusted to behave in the desired ways and must therefore be hedged about with rules, controls, checks, monitoring devices, or discipline.
> (1974: 37–38)

Conclusion

To conclude, the use of performance appraisals clearly represents an attempt to formalize a relationship between individual employees, their work and their managers in much the same way that project management systems tried to do the same for the relationship between the market and the work itself. Their presence indicates that management did not simply trust organizational professionals to work in the firm's interests. As such, they can be interpreted as a move towards the further rationalization of work relations and as evidence of the bureaucratization of authority within the work situation.

Even so, it should be stressed that in practice these mechanisms were of much less significance as a form of labour control than is sometimes portrayed (e.g., Savage *et al* 1988; Causer and Jones 1990; 1996). To reiterate a point made earlier, despite the fusion of technical and managerial roles, the previous technical experience of the managers and supervisors, and the use of bureaucratic monitoring mechanisms, management control was still far from complete. This, I argued, was because of the indeterminacy of technical work and because of management's own limitations.

The work of these employees was accordingly characterized by a relatively high degree of autonomy while the managerial role was one of technical facilitation rather than imposition. This fits with the general pattern of managerial authority found in case studies of technical workers in the US (Zussman 1985: 108–110), Britain (Whalley 1986a; Smith 1987: 127; Causer and Jones 1996) and France (Crawford 1989: 100–113). Furthermore, some of these studies have also found that junior engineers and technicians are more closely supervised than engineers (Whalley 1986a: 58–60; Crawford 1989: 100–113). In this context, a collegial style of management was inevitable in the case of the engineers and chemists. This can also be described as one of responsible autonomy where, according to Friedman (1977), employees who are central to long-term profitability are allowed a degree of discretion in their work on the understanding that management can trust them to use it in a responsible manner.

Unfortunately, the presence of powerful expectations tends to be forgotten in accounts of collegial as a style of management. These are important because they indicate that ideological control is present. What management cannot achieve through policy, they try to achieve through the shaping and maintaining of values, what I call a process of management by expectation. They tried, in the first instance, to ensure that the staff they recruited displayed a sense of professional pride (see acceptability criteria in Chapter 3). They subsequently drew on shared conceptions of professionalism and standards of work to ensure that these employees achieved a certain level of performance. Yet as 'the numbers game' demonstrated they could also be a source of conflict. What this conflict also revealed was that for all the emphasis on professional standards these were ultimately subservient to a conception of productivity that was governed by short-term costs and profit.

It follows that the mechanisms of bureaucratic control, such as performance appraisals, are best viewed as being embedded within, and lubricated by, a collegial pattern of employment relations. These relations influenced the way managers operated these policies since they knew that the nature of the work was such that it could not be specified precisely in advance. It should be remembered here that even the direct form of managerial control, that is, supervision, was incomplete despite the background of the various managers and supervisors. Furthermore, since much of the work of the engineers and chemists was based on projects the result was that they also held managerial responsibilities albeit of a fluid kind. The important point here is that this reinforced their inclination to act on the employers' behalf.

To conclude, the evidence presented in this chapter supports the claims advanced by Fox (1974) and Goldthorpe (1982; 1995) that the management of professional employment is conditioned by the employees' work orientation. Although these MNCs operated a range of bureaucratic monitoring mechanisms these were not the sole, or possibly even primary, source of employee performance. While they may not be quite so inappropriate as Fox (1974: 32) claimed, they were subsumed within a collegial style of management that exploited the technical enthusiasm and professional commitment of employees who would have found it difficult to obtain similar work in Ireland. In short, this analysis of the relations of authority in the employment relationship shows that the relationship is predominately one of service for engineers and chemists and partially one of service for technicians.

6

PAY, STATUS AND PERFORMANCE

Introduction

One of the key points of differentiation between the service and waged labour contract is the form of remuneration (Goldthorpe 1982; 1995). In the case of the former, the level of remuneration is closely associated with an employee's position on internal job hierarchies. Those who have worked their way to the top end of the job ladder receive a salary commensurate with their position. This contrasts with the situation of those in waged labour where levels of pay are linked to levels of effort. Hence they are frequently paid on an hourly or piece rate basis. This distinction is of special significance as the more open-ended nature of the effort reward bargain for those of salaried status is interpreted as evidence that such employees are trusted by their employers to perform as required. That is to say that the employer does not believe it necessary to introduce a direct link between levels of effort and reward in order to extract work from employees.

In the first chapter I outlined research which indicated that this distinction was inappropriate in the context of performance-based payment systems. Research by Savage *et al.* (1988) and by Causer and Jones (1990; 1996) have argued that the use of performance appraisals leads to informed evaluations by managers who have experience of working in technical roles. This leads to a significant individualization of reward in professional and managerial jobs: 'we would argue that it is important both in itself, and in opening up the way for a linkage of performance and rewards in a rather more immediate way than models of "high-trust" management might be taken to imply' (1996: 111–122). As I indicated in

Chapter 1, the significance of this for Savage *et al.* (1988: 463) is that: 'Employers do not simply "trust" that these workers will get on with it: they devise measures to ensure their actual work performance can be evaluated and their rewards calculated accordingly'.

The emergence of such practices is indicative of the increasing rationalization of the effort reward bargain. Under these practices white-collar employees working in large corporations are no longer paid a fixed salary based on status. Part of their salary is now at risk and can only be earned through a high level of performance. Furthermore they have to do so in a way that is transparent and meets the socially constructed criteria (Grint 1993) of whoever decides the merit-related element. This can be interpreted as part of a general trend towards McDonaldization where the drive for profit within capitalism has led to an ever increasing degree of predictability, calculability and control in social life (Ritzer 1993). This drive has incorporated some of these elements and in doing so would appear to have challenged the terms and conditions that go with white-collar employment status. This challenge has been further fuelled by the increasingly sophisticated management of the employment relationship through the use of HRM techniques (Smith 1992). Once introduced, performance-related pay and other policies are quickly established as 'best practice' within management circles. The attraction of such 'best practices' lies in the assumption that they eliminate slack, intensify individual effort and ultimately raise productivity.

The aim of this chapter is to investigate the extent to which management have succeeded in establishing a direct relationship between units of effort and units of reward for technical employees. This assumes that employers' remuneration strategies actually seek to link pay and performance in a significant way. But is this really the case? The answer to this will depend on what objectives lie behind the employers' overall approach to pay. Here the key issue is the extent to which employers have moved towards a merit-based payment system. There are a number of questions here. First, have employers completely discarded a system that is primarily based on hierarchy and professional status in favour of one that is based on individual performance? Second, if they have instead sought to combine the two, which approach dominates? In tackling these questions I shall place particular emphasis on recent changes in pay practices since these should reveal the path that these employers intend to follow.

Payment systems and employment status

Apart from Harford Laboratories, the payment systems used in these firms were corporate policies. Designed at the corporate headquarters, they were implemented in each division and branch plant across the globe. Harford Laboratories was unusual in that it had developed its own system of grades, merit assessment and bonuses following the establishment of a personnel department in the Irish subsidiary in the later 1970s. These were later brought into line with those of the corporation. The personnel departments were responsible for implementing and, if necessary, adjusting corporate-wide policies to suit local requirements. Part of their task was to compile an annual audit of local pay scales and submit it to the corporate personnel office. Despite this requirement the personnel managers in each of these factories insisted that their relationship with the corporate office was mostly one of communication and consultation rather than of rigid corporate control. However, they also admitted that local practices had to stay within certain specified limits. For example, pay increases had to fall within certain parameters set by the corporate office. This restriction was not one that local management resented because they were able to exploit it when bargaining with trade unions. Management frequently insisted that their 'final offer' was the best the unions could hope for as any further increases would not be sanctioned by the head office.

A striking feature of the payment systems was the clear distinction between those on salaries and those on wages. This division was such that employment status was actually defined by type of payment system: employees were categorized as either *salaried* or *hourly*. The slight exception here were unionized technicians in Kenine Pharmaceuticals and Harford Laboratories who were classed as 'semi-staff' status. In practice, the salaried included all those of white-collar status: managers, supervisors, engineers, chemists, the clerical staff – except for Harford where the clerical staff were unionized – and the technicians in Mertel Telecommunications (Figure 6.1). All of these received a proportion (less than 8 per cent) of their annual salary on the basis of their performance (Figure 6.1).

Firms	**Kenine**	**Harford**	**Mertel**	**Trojan**
Salaried status (white-collar, non-union technical)				
Time	Monthly	Monthly	Monthly	Monthly
Salary system	Hay–MSL system	Corporate system	Corporate system	Corporate system
Appraisal	Corporate system	Local system	Corporate system	Corporate system
– Frequency	12 months	12 months	3 months	3 months
– Percentage	0–6	0–8	0–7	0–7
Overtime	<450 points	Yes	Yes	Yes
Bonus	3 weeks	Merit	Length of service	No
Hourly status (blue-collar, unionised technicians)				
Time	Weekly	Weekly	Weekly	Weekly
Wage system	Collective bargaining	Collective bargaining	Collective bargaining	Corporate policy
Overtime	Yes	Yes	Yes	Yes
Bonus	3 weeks	2 weeks	Length of service	Attendance bonus

Figure 6.1 Employment status and payment systems
Note: The unionised technicians in Harford and in Kenine had semi-staff status

Salaried status

Apart from Kenine Pharmaceuticals, the salary structures in these companies consisted of a range of grades and increments for each position. These were arranged in an overlapping hierarchy of increments – each position ranged from 80 per cent to 120 per cent – so that the top increments for each position were worth slightly more than the bottom grade of the next position. For example, the top end of the senior technicians' scale was worth more than the bottom end of the scale for engineers.

Kenine Pharmaceuticals used the salary administration system devised by the management consultants, Hay–MSL. This ranked jobs by assigning a points value to each one in a quasi-mathematical system. Each job was evaluated along three criteria: technical sophistication ('know-how'), problem solving and accountability. Points were awarded for each of these categories to give an overall Hay-point level for each individual job. Salaries were then mapped onto the 'control point' (centre point) of each job using information from the annual Hay–MSL survey of other companies in the same industry. There was considerable variation between the companies in the frequency of appraisal meetings and consequently in the length of the review periods. To repeat a point made in the previous chapter, the appraisal meeting between manager and employee took place every year in Kenine and Harford, and every quarter in Trojan and Mertel. But in the latter two firms the technicians, engineers and even some managers admitted that there was a certain amount of 'drift' in the administration of the system: they only had one or two appraisal meetings in the entire year. When these performance review meetings eventually took place, a considerable amount of time was spent sorting out what the employee had actually worked on since the previous appraisal meeting.

Hourly status

The semi-skilled employees along with the technicians in Harford and Trojan were paid on a weekly basis. Apart from those in the non-union Trojan Electronics, none of these employees had any part of their pay related to individual performance. However, the management in Harford Laboratories would have liked to have had the technicians included in the appraisal system principally because it would have given them a greater capacity to link pay to performance and so increase productivity among the technicians.

One of the senior managers explained that it was 'something we have discussed off and on with the union and they're a bit anti about it to be honest'. The technicians' union, Manufacturing, Science and Finance, were strongly opposed to this because they were afraid that it would enable managers to identify and isolate those technicians who, in their subjective judgement, were 'poor performers'.

There were two further distinctions between those of hourly and salaried status. First, the 'hourlies' had to 'clock in' each morning and 'clock out' each evening. Second, those of hourly status had a lower annual leave entitlement. The technicians in Kenine Pharmaceuticals – informally labelled as 'semi-staff' – were differentiated from their SIPTU colleagues by being paid on a monthly basis, by not having to 'clock in' and also by having forty hours of leave which could be used in emergencies. The wage system for these employees was also based on a system of grades with annual increments. Wage increases were determined through collective bargaining. In Kenine, SIPTU negotiated for the general operatives and for the technicians. The resulting agreements were known as 'contracts', a term imported from US labour relations through corporate headquarters. Although these contracts were not enshrined in law (as in the US) they were, in practice, treated as binding within the Irish plant. They were generally negotiated on an annual basis though delays did occur. The union attributed these delays to management who they believed were trying to stall the negotiations by constantly referring the potential terms of any new contract back to the divisional headquarters in New Jersey for approval.

The technicians and clerical workers in Harford negotiated directly with the company through MSF while the manufacturing workers were represented by SIPTU. The wages settlements which MSF obtained were generally shaped by what SIPTU had agreed for the shopfloor workers. Management took care to ensure that any increases in the R&D budget for labour costs would be planned in conjunction with arrangements in the manufacturing facility. By doing this, they sought to avoid union claims for comparability, something that had created problems in the early years of the firm.

The management of remuneration

What then was the overall pay strategy of these employers? There are four dimensions to this question. The first concerns the structure of the payment system described above. The second is the position that these employers adopt within the labour market with regard to pay levels. The third concerns the way these employers managed the impact of collective bargaining on pay within these plants. The final dimension concerns the extent and significance of performance-related pay within the overall approach to remuneration.

Levels of pay

The four case study firms had reputations as 'good employers' within the local and national labour markets because they tended to offer generous reward packages: competitive levels of pay in addition to a comprehensive range of benefits. These included membership of company pension and life assurance schemes, sick pay schemes and subsidized membership of private health plans. The two chemistry-based firms were known to be among the top-paying employers in their industry while the electronics firms had more modest reputations. This was reflected in the survey data on levels of pay. Most of the respondents in the chemistry-based firms considered it to be very good or fairly good (Kenine 83.3 per cent; Harford 74.2 per cent) in comparison with other firms that they had knowledge of. Fewer of the respondents held the same view in the electronics firms. Two-thirds of the Mertel respondents (66.7 per cent) declared that pay was 'about average' while almost half of the Trojan technical staff (43.3 per cent) stated likewise. Even so, it was clear from the interviews that the electronics firms were still among the top-paying employers within their local labour markets. This tends to be a common pattern within the electronics industry (Geary 1991; 1992: 41).

The variation in employee perceptions of pay levels was not accidental. Its origins can be traced back to the employer decisions on labour market positioning. All four firms participated in industry-wide pay surveys and used this information to guide their own position. The personnel manager in Kenine stated that they adopted a position among 'the upper quartile of companies'. This decision was driven by their desire to recruit 'the best' technical graduates available in Ireland. Harford Laboratories also adopted a position among the top quartile of firms in the pharmaceutical and chemicals industry. They offered, in addition, an extra weighting for the Dublin region. This combination placed it among the highest paying employers within the entire industry. Mertel's pay strategy was to position itself just under the top quarter of firms in the electronics industry. While Trojan Electronics originally started off with the same strategy it had, according to the personnel area representative, changed its original strategy as a result of the corporation's financial difficulties. Its aim now was to pay 'the average of the market' as it could no longer afford large increases.

Pay bargaining: differentials and 'knock-ons'

In contrast to the firms studied by Savage *et al.* (1988) and Causer and Jones (1990; 1996) the pay practices of three of these branch plants were shaped by trade union policies and by established notions of status differentials. More importantly status differentials were still the central organizing principle behind the pay practices of these employers. Consequently, any attempt to establish a direct link between effort and performance operated within these boundaries.

The technicians in Kenine Pharmaceuticals and Harford Laboratories had in the past forced management to grant them pay rises that were so large as to have a distorting effect on existing status differentials. In Kenine, this increase widened the differential between the technicians and the general operatives. The general operatives were unhappy about the success of the technicians because they had broken the traditional differential and, what was worse, had done so by going alone. In subsequent years the personnel department in Kenine received repeated claims from the general operatives that were intended to restore the traditional levels of comparability. These were rejected repeatedly.

The technicians in Harford had taken advantage of the company's early surge in profitability to obtain large annual pay increases. They had also obtained relatively large pay increases and bonuses for operating new equipment. It was also 'custom and practice' for the technicians to submit an extra pay claim during the annual negotiations where there had been significant productivity increases as a result of certain changes in equipment and work practices. This was in addition to a lump sum paid at the end of each agreement period for allowing management to introduce whatever new technology or work practices they thought necessary. Despite recent efforts by Harford management to curb pay rises for the technicians, they still had to deal with the distortion in occupational differentials. The problem in this firm, unlike in Kenine, was with the salaried staff, especially the chemists and chemical engineers. They had become convinced that they were not being as well treated, in relative terms, as the technicians. In this context, it was not uncommon for newly appointed chemists to find themselves in a situation where they were delegating work to technicians who received more pay than they did! These status inconsistencies led to marked dissatisfaction among the chemists. Tom, one of the senior professional staff, expressed it in these terms:

> 'They are all madly overpaid anyway, the technicians in this place. That's changed a little now. There was a time when I had a technician working for me who was earning more than I was, which was crazy but I have since moved well ahead of him.'
>
> (Tom, senior chemical engineer, Harford Laboratories)

This status inconsistency in pay eventually caused such widespread discontent among the professional staff that they had discussions about forming a staff association and sent deputations to the vice president of R&D to discuss the problem.

Harford management responded to this problem by paying additional bonuses to the professional staff after the official pay increases. According to the Vice President of R&D a policy decision was taken to ensure that bonuses for the non-unionized staff were 'that bit bigger'. Company shares were also offered to the professional staff at a discount. At the same time managers began to stress to the chemists that their salaries would eventually overtake those of the technicians. In doing so, they emphasized the point that the professional staff should expect to have higher salaries because their responsibilities would ultimately be much greater than those of the technicians. As the R&D manager explained:

> 'From a chemist's viewpoint what we tell them is don't worry about what so and so . . . "He's been here twenty years, why shouldn't he have a good salary? In time you will earn more money." But the potential for the chemist is greater because ultimately his responsibilities will be greater, and it does work out that way.'
>
> (R&D section manager, Harford Laboratories)

A slightly different approach was taken in the management of these differentials in Mertel Telecommunications. It was company policy that the salary increase given to white-collar staff was always larger than that given to the shopfloor employees who were represented by a general union. But again, according to the personnel manager, it was the agreement with the trade union that decided 'what you give to the salaried people as well'.

This practice was common knowledge to the unions in these three firms and a source of annoyance to their members. Some resented the fact that they had to bargain and occasionally fall out with management in the annual negotiations before they could get a pay rise. Yet whatever they got would then be handed automatically to the salaried staff along with an extra percentage or a bonus increase.

Towards performance-related pay?

How far did these employers succeed in linking pay to performance for those on salaried status? One possible insight into management's objectives was the proportion of pay that is related to merit (Figure 6.1). Another was the nature of changes made to the system by management.

Each of these firms used the same percentage increase formula for performance-related pay. This was typically added to a general cost-of-living increase. However, the actual percentage was at its largest (Harford Laboratories) still only 8 per cent. The size of this increase for individual employees was decided by their manager on the basis of an appraisal rating. This in turn was part of a gross salary which, in the Irish Republic, was then subject to a personal income tax of 48 per cent. It would not, therefore, seem appropriate to say these payment systems contained a significant performance-related element. Nor was it seen as such by managers or employees. If anything, these systems were based on merit at the margins rather than on any significant individualization of reward.

It might be assumed that any changes these employers made to the pay system would introduce a greater degree of individualization of reward. This could be achieved by simply increasing the proportion that was based on merit. Such changes would also be in line with trends in human resource management that stress the link between market-based organizational goals and individual goals and between pay and performance (Kessler and Purcell 1992: 16). They would also fit the general trend away from simply trusting these employees (Savage *et al.* 1988). This, of course, assumes that individualized performance-related pay was a management objective. At the time of this research three of these employers, Trojan, Kenine and Mertel, were in the process of changing the payment systems for their salaried staff. This provided a unique research opportunity to investigate the objectives that management had for their reward strategies.

In two of these cases, management were trying to strengthen the relationship between pay and performance. However, these changes were only marginal adjustments. Furthermore, the changes that came about in all three branch plants were not triggered by a decision to expand the performance element of the pay package. The impetus came from other sources.

The pressure for change came from two different directions in Trojan Electronics. The first was the widespread dissatisfaction with the operation of the existing performance appraisal system among the workforce. The personnel department was also unhappy with the continuing 'upward drift' in the ratings because this raised labour costs. In the first few years it 'was driven hard to have 60 per cent in the Fully Satisfactory rating, but the pressure went off'. When the personnel department audited the results they found that almost 70 per cent were in the Excellent category. As part of their review the personnel department organized focus group sessions with representatives from different departments. The conclusion was that there was too much variation in the interpretations of the supervisors and managers as to what constituted 'Excellent' or 'Fully Satisfactory' rather than there being any fault in the structure of the system itself. They recommended that further training sessions be provided for supervisors and managers with the emphasis on 'hardening the line' pending the introduction of the new three-rating version of the review system. This new version, which reduced the number of ratings from five to three was already in use throughout the rest of the corporation. The Irish plant was under pressure from corporate headquarters to implement the changes and come into line with its sister plants.

In Mertel Telecommunications the changes to the MFA and salary systems came from the same two sources as in Trojan. Mertel's salary grading system was changed as part of a corporate-wide procedure. In the old corporate system there were fifty-four different salary grades. The senior management in the Irish plant held the view that this was too cumbersome and inhibited flexibility. The fifty-four salary grades were, consequently, being reduced to thirteen 'bands' to allow for greater salary variations within these broader ranges. There was some uncertainty over what effect this would have in practice though the personnel manager hoped it would lead to greater flexibility between departments. The 'Managing For

Achievement' system of appraisal was changed primarily because of the level of criticism that it provoked among the salaried staff. Prior to this change the various department managers came together to assess the overall balance of ratings within the factory and to contribute to the ratings made by managers in other departments. This latter aspect of these meetings had to be discarded because those being appraised did not wish to have their reputations tossed backwards and forwards among managers for whom they did not work.

In Kenine Pharmaceuticals, the reason for changing the salary system was that the more senior chemists and chemical engineers were reaching the maximum of their grades within the system (i.e., 120 per cent). The personnel manager held the view that those employees had limited financial incentive to continue to give of their best efforts. The change would incorporate the 'cost-of-living' increase into the merit rating rather than giving it automatically each year. Traditionally, the cost-of-living increases had been kept separate from the merit-based increase. The change meant that this particular increase could no longer be expected, it too had to be earned. The personnel manager explained that this probably implied that there was more of a 'stick than a carrot' introduced with this change:

> 'I think it's sufficient incentive when you say to a person "Well look if you want to stay on your max then your performance has got to be up where it always has been, when you got up there, and you cannot have any slippage in your performance." I think that's incentive enough. In this new system there is probably more of a stick than a carrot.'
>
> (personnel manager, Kenine Pharmaceuticals)

To conclude, none of the changes in these firms were designed solely to accentuate the relationship between performance and pay. Those changes that did come about were either as a result of corporate-wide initiatives or as a result of widespread dissatisfaction with aspects of the existing policies at local level. This does not fit with much of the received wisdom on human resource management which assumes that employers should seek to define and reward individual performance (Beer *et al.* 1985; Guest 1987). The conclusion must be that prescriptive models of HRM fail to account for the limits imposed by organizational politics.

Pay relating to performance

For most of this chapter I have argued that employers were not primarily concerned with introducing individualized performance-related pay of a significant kind. Instead they sought to maintain competitive salary levels within their respective industries; to maintain differentials between different status groups within the firm; and to bring their appraisal systems into line with those of their respective corporations. However, in this final section I shall show how these employers did offer a form of merit-based pay. In contrast to the arguments advanced by Savage *et al.* (1988) and indeed the best practice HRM literature, these awards were of a traditional nature. These, I argue, reinforced rather than eroded the traditional elements of professional employment. Two such practices will be described: the rapid progression of graduates through the initial pay grades, and the distribution of discretionary bonuses to 'key contributors'.

Each of these companies had a policy of quickly pushing newly arrived graduate engineers and chemists up through the initial salary grades. This was colloquially termed 'the fast track'. The initial salary level for graduate engineers and chemists was relatively low compared to other technical staff because they had limited full-time industrial experience. The 'fast track' was a response to the rapid early development of these graduates. Salary level increased in a direct relationship with the individual's ability to move along the early learning curve. The management argument was that fresh graduates were of little practical use for the first eighteen months. Consequently, their pay could not be very high during their early years in the organization. The rationale was that these new recruits were learning at a rapid rate so they could therefore be given some larger rewards. One engineer described this process as follows:

> 'If you take a graduate and you leave him as a graduate engineer for a few years he is going to get a bit despondent and he'll probably leave after a couple of years. Particularly with graduates, they do need to have reviews on a fairly regular basis because they are developing fairly fast and they are developing skills and they need to see that those are being rewarded.'
>
> (Dermot, senior test engineer, Trojan Electronics)

One manager explained that in the early years of an employee's work history this practice was effective because money was important to them at that stage:

> 'So we can do a lot of motivating for quite a number of years just through the administration of the salary structures. And if a person is a consistently high performer that person has the potential to move to 115 per cent of his nominal salary range. As well as that then there is, particularly in the graduate group, scope to upgrade the person's grading as the person progresses. There's really a lot of scope for a person even within his own job to progress.'
>
> (training and recruitment manager, Kenine Pharmaceuticals)

The training and development manager's argument was that this process of pushing newly arrived staff up through the grades provided substantial motivation in itself particularly for the better 'performers'. Regular increases through the early steps of the salary system provided definite positive feedback of career progression. The 'fast track' was also used to reduce feelings of status inconsistency in Kenine and Harford where the technicians earned more than the graduate chemists and chemical engineers.

The rewards to those who were seen to be outstanding performers were of two kinds: the formal and the informal. Harford and Mertel operated special annual award schemes for employees who made exceptional contributions. In Harford the award was a corporate-wide scheme designed to reward outstanding scientific work. Named after the founder of the company, 'The Krieble Award' was perceived as being very prestigious though it also had a significant financial aspect. Some of the senior managers in Harford R&D had won the award earlier in their careers – typically for inventions that made a lot of money.

In Mertel there were two such awards though they were limited to the local branch plants. The first was a 'Cost Reduction' scheme. Any employee who came up with an idea to save the company money in, for example, the design or manufacture of a product received 5 per cent of the savings in cash. The only restriction was that the idea must relate to something that was not part of the employee's own job. The second scheme, the 'Recognition Award', was designed to bring formal recognition to the efforts of individual employees. Employees could submit detailed nominations to the

Recognition Committee (consisting of managers drawn from different departments) on behalf of others or for a group of employees whom they believed had made a significant contribution to the firm. The awards were gold chains for women and tie pins for men.

The awards allocated in the chemistry-based firms were cash bonuses which were totally at the discretion of management. Some, like the 'once-off payments', were not officially announced or acknowledged. These, according to a past recipient, were 'a very private type of thing between you and your boss'. They tended to be given to employees who had made 'key contributions' to the firm and whom the firm were keen to retain. Since these bonuses were not always available or well publicized, and since they were distributed sporadically, they did not act as an incentive.

Bonuses were also occasionally distributed in Kenine and Harford if the firm had an exceptionally profitable year:

> 'Well there is a Christmas bonus which everybody gets. It is a standard thing. In the last few years there have been kind of performance-related bonuses – once-off lump payments – I think it depends on how the whole company performs. And if there is a good company performance, you know financial performance at the end of the year, I presume there is a proportion allocated to the people who contributed to achieving that.'
>
> (chemical engineering manager, technical operations, Kenine Pharmaceuticals)

Inventions in Harford which were registered as patents could lead to royalties as well as patent *honaria*. The company also paid £200–£300 for every patent registered. These tended to be tax free. Even so, many of these awards and payments represented a pecuniary acknowledgement rather than a financial attraction.

Finally, Harford operated a Christmas bonus scheme for the technicians. This was management's sole mechanism for distinguishing between the performances of the technicians. It comprised a plus or minus 10 per cent bonus calculated from two weeks' pay. It was a bitterly controversial practice that fostered a great deal of resentment among those technicians who were deemed to have performed less well. Part of the reason for this was that judgement was based on the views of the departmental managers and chemists who acted as supervisors in R&D rather than on any formal system of appraisal. The amount of money involved was not significant, usually around £30 or £40. Still as one of the technical specialists explained: 'It's not actually the amount of money or anything it's more the principle. "This is your disapproved bit for this year"' (Marie, technical specialist, R&D Harford Laboratories). The technician's union had tried on a number of occasions to get the bonus scheme changed to a single 'flat' bonus for all. They made no progress. Management, and the R&D managers in particular, would not agree as they felt that the principle of distinguishing between technicians was all important.

The range of bonuses and awards described here corresponds with those found in other studies of reward strategies for R&D staff in high technology firms (Gomez-Mejia and Balkin 1989). What is remarkable here is that even in the context of sophisticated hi-tech work environments one of the more significant links between pay and performance was not determined by a rationalized bureaucratic process. Instead it was made by a rather traditional method: management discretion.

Conclusion

The evidence presented in this chapter has shown that employers' remuneration practices were primarily concerned with the maintenance of competitive levels of pay, status differentials and seniority-based pay rather than with the development of individual performance-related rewards. While there was an element of performance-related pay within the various payment systems this was not the guiding principle. The payment systems were instead organized around clear status differences between the *salaried* and the *hourly* employees. These differences produced pay differentials that were consciously maintained by these employers.

An interesting feature of these status differentials is the way they contrast with those described in an early study of industrial technicians within British industry (Roberts *et al.* 1972). Roberts *et al.* found that unionized manual workers were closing the traditional gap between their wages and the salaries of unorganized technicians. This, they suggested, was one of the main reasons why technicians were joining trade unions. In the case of two of the firms studied here, the problem was rather different in that it was the wage gains of the technicians that caused the problem. It was the chemists in Harford and the process operators in Kenine who were dissatisfied with the changes in the traditional pay differentials. This does not necessarily contradict those of Roberts *et al.* (1972) since it could be argued that it actually confirms the benefits of trade union representation for technicians. Rather than being threatened by the rise of manual workers' pay, it was the technicians who threatened those further up the job hierarchy.

In relation to the overall argument, the key finding is that the division between service and waged forms of labour was maintained in the form of a division between the salaried and the hourly. The hourly employees also had to clock in as well as being directly supervised while at work. In short, Weber (1948) and Goldthorpe's (1982) distinction between those who earned salaries and those who earned wages still holds. Yet the presence of a performance-related pay policy for the salaried staff would seem to indicate that these employers do not simply trust their staff. First, the application of such practices as performance appraisals has obviously meant that the employment relationship for service class members is no longer simply characterized by a 'code of service'. Instead it seems, at least on the surface, closer to the traditional 'labour contract' of wage workers where units of work are measured and related to pay scales. The irony here is that those on the salaried system would seem to have more formalized bureaucratic controls governing their work situation than the hourly unionized workers.

It would, however, be a mistake to regard the presence of such policies as a fundamental change in the nature of professional white-collar employment. The proportion of pay involved was no more than 8 per cent and, more importantly, the changes made to the payment systems did not seek to extend this. If anything, the payments that were of significance actually reinforced the traditional conceptions of service-class employment. The bonuses, in particular, were, in effect, a form of compensation for service rendered. This is particularly true for those of the informal kind. Furthermore, such payments were at the discretion of management and, as such, could not be assumed to be available for each extra special effort. Similarly, the use of the graduate fast track for the professional staff emphasized experience and seniority. These payments were also used to reinforce status for those who earned less than technicians. In other words, they reinforced one of the traditional elements of service-class employment.

7

CAREERS, LABOUR MARKETS AND JOB HIERARCHIES

Introduction

The traditional sociological view of the white-collar career within an organizational setting is that it is characterized by regular, incremental increases in authority, status and salary (Weber 1948). As I indicated in Chapter 1, this view dominates Goldthorpe's (1982) conception of the service-class employment as one where employees have an understanding that their service will be rewarded with job security and career opportunities combined with increases in salary. It is also one that is supported by numerous studies of managerial employment within large organizations (e.g., Kanter 1977; Rosenbaum 1984; Nicholson and West 1988; Roomkin 1989). Similar findings have appeared in studies of various professionals employed in industrial organizations such as scientists (Kornhauser 1962; Marcson 1969).

The persistence of these arrangments can be attributed to the return that the employer receives for providing a long-term investment in human capital. Employers have a clear incentive to develop what economists call internal labour markets when their business requires the accumulation of firm-specific knowledge. An internal labour market is defined by Doeringer and Piore (1971:1–2) as:

> an administrative unit, such as a manufacturing plant, within which the pricing and allocation of labour is governed by a set of administrative rules and procedures. The internal labour market, governed by administrative rules, is to be distinguished from the external labour market of conventional economic theory, where pricing, allocating, and training decisions are controlled directly by economic variables.

The provision of a career within an internal labour market allows specialist employees, such as engineers, time to fine tune the application of their academic knowledge to the specific requirements of their employer and *vice versa*. The result is that the longer these employees stay with the same employer the greater their contribution.

Despite these advantages, there is an accumulating body of evidence that indicates that this model of white-collar employment is undergoing significant change. Managerial career paths have, as indicated in Chapter 1, been delayered by recent waves of corporate downsizing, notably in the US (Hirsch 1993; Heckscher 1995; Osterman 1996) and in the UK (Brown and Scase 1994; McGovern *et al.* 1998). Some of these changes have also been experienced by organizational professionals. Whittington's (1990; 1991a; 1991b) research on R&D staff in fourteen UK organizations found that restructuring measures had destabilized their job security and status as 'core employees'. This was because R&D was no longer viewed as a core activity that firms had to insulate from market pressures. Some of these activities could instead be 'contracted out'. As I also indicated in Chapter 1, Savage *et al.* (1992) view these changes as part of a general process in which bureaucratic hierarchies are becoming less significant in economic organizations. Thus, the decline of the internally promoted manager can be explained by a marked tendency for employers to use market mechanisms rather than job hierarchies to structure their activities. They also argue that professionals fare better in this environment because they possess academic qualifications.

These developments raise specific questions about employer practices which can only be addressed properly at the organizational level. If there is a move from the internal labour market towards the external market as a means of regulating white-collar employment, then this should be evident in the practices of individual employers. It is, however, by no means certain that large numbers of employers are making this shift. Those who advocate best practice models of HRM argue that the development of career paths is essential for hi-tech employers precisely because they must attract and retain knowledge workers. Again there is empirical evidence to indicate that such practices do exist. Some US hi-tech employers, for example, have integrated programmes for career development into their performance appraisal systems. This enables managers to assess the employees' strengths and weaknesses, their needs for further training and possible career options (Miljus and Smith

1987). Other evidence outside the HRM literature indicates that employers are prepared to increase promotion opportunities for engineers by inflating titles (Zussman 1985: 141–142). Some employers create new management positions in an effort to prevent them from leaving (Whalley 1986a: 97). Large-scale 'organizational fixes', such as the creation of dual career ladders, have also been introduced (Gunz 1980).

The problem with this research is that it is situated in national labour markets that are characterized by skill shortages. This then shapes the employers' behaviour and the way it is interpreted. However, the major weakness is that this experience is assumed to be universal although specific to hi-tech industry. This still leaves the question of how employers behave when they experience a skill surplus. Is it safe to assume that MNC employers will have similar policies in the Irish context? Did they seek to convert, what Althauser and Kalleberg (1981) call, the occupational internal labour market (OILM) of the engineers and chemists and the occupational labour market (OLM) of the technicians into a firm internal labour market (FILM)? Furthermore, are such policies possible within MNC branch plants if, as the NIDL thesis suggest, they do not undertake technical work of sufficient sophistication to warrant the development of dual career paths? How then do these employers handle the aspirations of these employees?

This chapter develops the discussion of labour markets and employer strategies that began in Chapter 3. It begins by describing how the desire to learn new knowledge and to make a reputation motivated these employees during the early part of their careers. It then examines how these employers dealt with the desire for promotion among staff who had advanced beyond the early career stage. The final section explores whether the practices of these employers were conditioned by the prevailing labour market conditions. The key question overall is whether management strove to retain existing employees or alternatively managed a turnover of staff in order to recruit 'fresh blood'.

Knowledge, projects and promotions

The importance that these employees placed on their career needs was reflected in a number of ways. First of all, it was influential in the decision to join the firm. Indeed, there is considerable evidence to suggest that the prospect of future career opportunities may be as important for engineers as the salaries offered by potential employers (Bailyn and Lynch 1983; Steiner and Farr 1986). In the short to medium term the MNCs succeeded in meeting some of the career needs of their technical staff through the work that they provided. One such need was to learn more about chemistry or electronics. Another need was the provision of 'quality work' (see Chapter 5) which extended an individual's technical experience and boosted their prospects for promotion.

New knowledge

From the interviews with these employees, it was clear that they had a strong desire to learn as much they could about their jobs, about the products that they worked on and the relevant areas of chemistry or electronics. Most were still in the early stages of their careers and were simply enjoying their first job after college. As the following comments indicate these employers offered opportunities to work on state of the art equipment, on interesting projects and on new technologies:

> 'At the moment I enjoy my work, the work I have is challenging and it's technically challenging and I'm confidently growing and improving and I can see that. . . I've also had a very varied career so far. My first two years I was twelve months off site in other plants and this is the first year I have actually spent on site.'
>
> (Niall, chemical engineer, technical operations, Kenine Pharmaceuticals)

> 'It has to be said, in my situation, this is a very interesting place to work. It really is. Projects change and you don't end up working in the same thing for too long really. You get new equipment coming in, sophisticated equipment, computerized equipment, etc. There are always new things to be learned.'
>
> (Tom, senior chemical engineer, Harford Laboratories)

> 'I have learned more in the last twelve months than I'd have learned in ten years in the other place [previous employer], simply by exposure. If I never did a training course by the exposure to the equipment, that in itself is a great help.'
>
> (Alan, technician, Trojan Electronics)

A desire to develop their CV was also present as a significant proportion of these employees expected to change employers once they had learned all there was to learn from their current employer. They had few personal restrictions and were highly mobile as almost two-thirds of these employees were less than thirty years of age and had not formed a family unit with dependants. It was not therefore surprising to find that almost half (45.7 per cent) of the survey respondents were taking, or had taken, part-time degree courses (mostly BScs for technicians and MScs for the engineers and chemists). Most of these (61.5 per cent) were motivated by career development reasons. Also, of the quarter or so respondents who were members of a professional association, almost half (48.5 per cent) indicated that they saw their membership as a way of obtaining technical information. Individual members, notably in Harford, explained that their membership allowed them to keep up to date on technical developments and also gave them a certain professional status that they could use when changing employers.

'Quality work' was also highly valued for what it added to an employee's progress within his job and within the firm. This, as already described, was used by individual managers to motivate their staff though their ability to do so was constrained (Chapter 5). Project experience was of particular importance, especially for the engineers and chemists who were interested in moving into management.

Project experience and reputation

Along with the search for new knowledge, the early years were also marked by a search for work opportunities that would help to establish a reputation. The generation of a reputation was important in increasing promotion prospects and developing a long-term career within the plant. An engineer, for example, who developed a strong reputation in his/her early years might be expected to gain a number of promotions before eventually joining managerial ranks. This process required a considerable element of self-promotion. Engineers and chemists learned to be 'pushy' during their formative years with their first employers. One engineer in the technology unit in Mertel explained: 'after a while you realize that you have to get in there and kick up a bit of a stink and to battle more for what you want'. One such practice was to make job performance visible to managers. In this way an employee got his technical abilities noticed and gradually began to build a reputation. One manager explained the process in this way:

> 'A lot of the engineers would look around to find projects that would give them greater visibility externally and internally so they are seen to be high performers. So when a position does arise immediately you do look at this guy and he goes for it.'
>
> (director of quality control, Mertel Telecommunications)

Certain types of technical work were highly valued for the levels of responsibility and visibility that went with them. The best opportunities came with participation in major projects, especially as project manager. The more high profile the project, the greater its capacity for exposure, and consequently the more it was prized. High-profile projects were those which would have a major impact on the operation of the plant or on its market share. One example was the introduction of a more efficient production method for a specific product line. Such projects could have a major bearing on the future of the plant itself so the responsibility associated with them could hardly be much higher. The end of a project report would bear the nature of the project leader and the various team members. In such instances the name of the project manager became well known within the local plant and possibly within the corporation.

For example, one of the chemical engineers (Vincent) in the technical operations unit at Kenine Pharmaceuticals was leading a project team that would introduce a new chemical process. Considerable periods of time were allocated for the project in the production schedule even though this was at great expense to the Irish plant. Should it be successful then the Irish-based plant would be among the most cost efficient in the corporation's manufacturing division. The priority attached to it meant that it was very 'high profile'. Vincent's role was to coordinate the implementation process with the materials manager, with the quality control department, and with manufacturing projects. He had to ensure that the appropriate equipment was procured from the research laboratories in the US. Consequently, his name became widely known within the corporation as well as within the Irish plant. This, he explained, would unquestionably be of benefit to his position within Kenine: 'The more you are in the limelight the better chance you are going to get promoted or a salary increase or you get an award for what you have done.'

This was also common to staff in the other firms, with some variations. For instance in Mertel the engineers in the technology unit had to deal directly with customers when making presentations on the capacities of new products or when installing new PABXs for a customer firm. On such occasions, customers might contact the company to express their satisfaction with the work which that particular engineer had carried out for them.

The processes of generating a reputation ensured that employees performed in a responsible way for the employer because of two distinct pressures (Burawoy and Wright, 1990: 262). First, it involved the positive demonstration of an individual's achievements through self-promotion: individual employees monitored their own performances in a way that was of direct benefit both to themselves and to their employer. Second, a reputation could only be translated into promotion if it was publicly recognizable. Together, these practices served as a form of self-induced social control in technical work.

Promotions and career paths

Managers were only too aware of the preoccupation of these employees with obtaining promotion. The personnel area representative in Trojan Electronics put it quite frankly: 'Promotion, I would say, is the be all and end all of every technician and every engineer'. He also emphasized that promotion was part of a deeper need for recognition among the technicians. Their main contention was that they wanted more recognition for their contributions, principally in the form of promotion to associate engineer. Another manager explained that he and other managers had to discuss promotion prospects directly with engineers even to the point where it might have to be admitted that opportunities might not arise in Mertel:

> 'A lot of engineers are career orientated. They will settle for 12 to 18 months and then they want to see where they are going in the long term. They would not be prepared to sit into a job, to sit there year in year out, so we give them feedback where we think they are capable of promotion and if the opportunity arises and we say: "You have the ability to move onto the next level". But that opportunity might not arise in this company.'
>
> (director of quality control, Mertel Telecommunications)

Were these employers able to meet the long-term career needs of these employees? The survey data on the availability of promotion opportunities suggests that they were not (Table 7.1). The results from all four firms revealed that three-quarters of the respondents (74.2 per cent) believed that promotion possibilities were either fairly poor or very poor for employees at their level. The two electronics firms represented the extreme cases. Only little over half (56.7 per cent) of the Mertel respondents saw promotion possibilities to be fairly, or very poor, while almost all of the Trojan respondents (91.9 per cent) did so.

The Mertel results were supported by a perception among the staff that there was a large amount of turnover among engineers. This in turn reflected Mertel's pay policies and the difficulties which young engineers faced in obtaining work experience that they saw as of benefit to their careers. In Trojan the results are best understood in the context of its recent history of job losses and delayering. Some promotion positions simply did not exist after the redundancies. With the possible exception of Harford, the longest established company, the respondents in the others may also have had expectations of promotion from the early growth phase. The initial expansion of these organizations meant that more and more opportunities became available in the early years as staff grew with the company. This was most noticeable in Trojan, the most recently established of the branch plants, where there were still memories of the rapid expansion that occurred during the initial start-up phase. According to the quality control manager, one-third of the employees had obtained promotion about once every three years. This practice had continued until approximately eighteen months

Table 7.1 Promotion possibilities at your level

Company	*Kenine* %	*Harford* %	*Mertel* %	*Trojan* %	*Total* %
Very good or fairly good	5.6	6.5	6.7	0.0	4.3
About average	27.8	16.1	33.3	8.1	19.8
Fairly poor or very poor	61.0	77.4	56.7	91.9	74.2
No response	5.6	0.0	3.3	0.0	1.7
Total	100.0	100.0	100.0	100.0	100.0
N=	18	31	30	37	116

Note: Number of categories collapsed from five to three

before the research began. This he believed had left employees somewhat spoiled, something that now had to be changed.

In both Mertel and Kenine the period of initial expansion had ceased a number of years earlier with a concomitant decline in the number of promotions available. The only options open to the chemists and engineers in these companies was promotion into management positions whenever such became available. Interviews in Mertel with the director of manufacturing, the systems design manager and in Kenine with the chemistry section manager and personnel manager confirmed that this had resulted in human resource problems.

> 'Some of the activity we do isn't on the leading edge. You have fairly ambitious, fairly motivated people and if you abuse that for too long, leave them doing things that eventually become reasonably mundane, they become demotivated. It's been a reason why we have lost some people.'
>
> (systems design manager, technology, Mertel Telecommunications)

The core problem was one of stagnation among the more experienced technical staff. They had become too comfortable to leave yet were not as productive as they had been during their earlier years in the firm. This problem was compounded by the inability of these firms to provide dual career ladders in the absence of the more sophisticated basic research and design activities. Except for Harford Laboratories, these branch plants were unable to provide extended scientific, or engineering, job ladders of directly comparable status to middle and senior management positions. Even when they took on more complex work they were still only in a position to consider the introduction of a *single* additional step at the top of the existing technical hierarchy. This 'ceiling' on the technical expertise within these plants meant that the technical career path was somewhat truncated. Apart from Harford Laboratories the job ladders for engineers, or chemists, only allowed for a maximum of three promotions (Figure 7.1). This meant that a graduate engineer (or graduate chemist in Kenine), for example, could initially hope to be promoted to 'engineer status' (chemist) before reaching the technical ceiling of 'senior engineer' (or chemist).

One direct result of the career structures was that 'career success' was ultimately seen as promotion into management (see also Zussman 1985; EOLAS 1989: 91). For instance, all of the technical

Kenine						
Technician	>	Senior technician				
Chemist		Senior chemist				
Chemical engineer		Senior chemical engineer				
Harford						
Technician	>	Senior technician	>	Technical specialist		
Chemist		*Chemist		Senior *chemist		
Chemical engineer		*Chemical engineer		Senior *chemical engineer		
*research, analytical or development						
Mertel						
Technician	>	Senior technician	>	Technical specialist		
Graduate engineer		Engineer		Senior engineer		
Trojan						
Technician 1	>	Technician 2	>	Technician 3	>	Technician 4
Graduate engineer		Engineer		Senior engineer		
Associate engineer		Engineer		Senior engineer		

Figure 7.1 Technical career structures

managers who were interviewed for this research had 'graduated from the bench', that is from positions as engineers or chemists. This pattern of progression only served to reinforce the importance attached to getting managerial experience through project management and technician supervision. This meant that, for some of the chemists and engineers, the top end of the OILM in these organizations merged with the firm's internal labour market. The options for others were either to remain in the same position or leave for a position elsewhere. This led to a significant degree of staff turnover which, as I shall argue later, enabled these employers to exploit the external labour market to meet their staffing needs.

The number of grades for technicians ranged from two in Kenine Pharmaceuticals to four in Trojan Electronics (Figure 7.1). In Harford Laboratories the technicians' ladder had three steps with the additional promotion-based position of technical specialist. Mertel Telecommunications also had a technical specialist grade and there were four technician grades in Trojan Electronics. In Trojan Electronics it was possible for technicians to continue to associate engineer status after 'tech 4' and, theoretically, to 'full engineer' status. However, the possibility for promotion out of their occupational labour market into the position of engineer or chemist had disappeared with the general rise in educational standards.

Technicians and the graduate barrier

The rising standards that these employers applied to educational qualifications had a dramatic impact on the internal labour market. In the past, when there was a prolonged skill shortage, it had been possible for those who entered the organization as technicians to work their way up the hierarchy to the position of chemist or engineer. This route included an intermediate position of 'technical specialist' (in Harford Laboratories and Mertel Telecommunications) or 'associate engineer' (Trojan Electronics). In recent years, however, the transition to the actual position of engineer or chemist had become blocked by the gradual introduction of a new criterion by management, namely, the possession of an undergraduate degree within the relevant discipline.

While it could be argued that qualifications indicated academic ability and trainability in the selection of new employees, such an argument could not be made for promotion: the candidates had already received training and some amount of work experience within the firm. Even so, each of these employers still insisted on the 'piece of paper':

> 'Basically the minimum requirement, for instance into engineering would be a chemical engineering degree. . . . In terms of going from being a lab tech to a chemist again the basic requirement is you must be a chemist which requires having a degree in chemistry.'
>
> (employee relations manager, Kenine Pharmaceuticals)

> 'A person going in with a diploma wouldn't be promoted to a chemist or an engineer. Sometimes if you go on and converted your diploma into a degree you may be promoted, you may just be promoted to technical specialist.'
>
> (Anna, senior technician, Harford Laboratories)

> 'The big difference is the piece of paper. That is all it is I guarantee that. I can give you a classic example where I went to the job advertisement board once for a component engineer and I went in for it and I went through all the interviews, was asked serious questions, got them all correct so on – the life of the capacitor, got all that right – and at the end of it the director here said and do you have a degree and I said you know I don't and he said "I am sorry we can't offer you the job".'
>
> (Stephen, technical specialist, technology, Mertel Telecommunications)

Even in Trojan Electronics, where the position of associate engineer was designed to allow technicians to become in-house engineers, there was a strong emphasis on qualifications. Technicians who wished to become associate engineers were interviewed by the Engineering Review Board. This committee of technical managers was established specifically to ensure that the standards of engineering were maintained when appointing associate engineers. The Board's principal means of maintaining standards was to interview candidates on their technical knowledge. If the candidates were successful they then had to undertake a specially assigned project over a period of two years before they were given the status of associate engineer. It was widely understood among the technicians that they would not be interviewed by the Engineering Review Board unless they had obtained a nationally recognized diploma. A senior technician explained there was an informal ranking of qualifications within the firm. This coincided with that of the higher education institutions that granted them:

> 'If you've got an NCEA Diploma you're on the pig's back, if you are reasonably good you will get through it. I think if you have got a Part Two in the City and Guilds it's a little harder, if you have got the in-house technician course then it is harder again. That is well recognized informally in here.'
>
> (Des, technician 3, production, Trojan Electronics)

The possibilities for promotion to technical specialist in Harford were also quite remote as the company believed that it only needed four 'super-technicians'. It was this predicament that led to the use of the term 'lifers' in Harford for those older technicians who had no possibility of promotion but were financially bound to the firm. One research scientist compared the situation to the worst aspects of civil service employment:

> 'They are at the top of their scales and that is a problem for us. How do we keep technicians interested if they don't have promotions? That has always been a difficult task to achieve from a management point of view. It tends to become a little bit like the civil service, you are promoted to the top of the ladder and you are stuck there and you are young.'
>
> (Conor, senior scientist, R&D Harford Laboratories)

The only option open to management, according to the vice president of R&D, was to 'get the group shunted around into new jobs to get them involved in projects'. Even this was not easy.

To circumvent this barrier many technicians were actively seeking degrees through part-time evening courses. Half of the technicians (50.7 per cent) either had taken, or were in the process of taking, part-time evening courses in order to improve their career prospects. Some of these were actually taking a degree through distance-learning in the hope that it would either help them to get over the graduate barrier within their current organization or help them to find employment elsewhere. Despite these efforts managers in Harford Laboratories and in Mertel Telecommunications had adopted a policy which stated that these qualifications were not sufficient. Their view was that they were not prepared to 'hand out' promotions to individual employees who had recently obtained degrees simply because they had stated in the past that engineers and chemists were deemed to be of degree status. In the two firms in which the technicians were unionized, Kenine Pharmaceuticals and Harford Laboratories, managers tended to point to the simple 'market reality' of the higher wage costs associated with technicians. The trade unions that represented the technicians in both of these firms had been very successful in winning large pay increases for their members in the past. This had reached a point where significant status inconsistencies had emerged between the chemists and the technicians where the more experienced technicians had moved up through the union negotiated seniority-based wage scales. Consequently, management preferred to recruit graduates from the external labour market rather than promote internally:

> 'An aspiring technician who, let's even say with three years of additional study over and above the NCEA Certificate, achieves, let's call it a BSc in chemistry. They're nobody, they're competing against Ph.D. chemists. Now from a pure recruiters point of view why bother to go and hire a BSc when you can go and buy an MSc or a Ph.D. for half the price. Now that is a total market reality.'
>
> (personnel manager, Harford Laboratories)

In addition to the higher wage costs involved in promoting a technician, the extra qualification that the technician had obtained would still (as the quote from this personnel manager indicates) leave them less qualified than what was available in the external labour market. From the management point of view, it made much more sense to recruit a Ph.D. chemist from university. This was because there was a 'big gap' between the knowledge contained in an undergraduate degree and in a Ph.D.:

> 'I think the other problem is that effectively, they would come from a technician's background, get a degree and then find themselves competing with a Ph.D. organic chemist. That's a big gap. If somebody gets a qualification as a chemist we would not automatically say 'OK now you are a chemist' because we can go out and get a Ph.D. chemist.'
>
> (section manager, R&D, Harford Laboratories)

Though the technicians were not unionized in Mertel Telecommunications, their managers also referred to the costs associated with promoting technicians whose salaries were higher than that of newly qualified engineers emerging from university. The argument concerning the gap between a basic degree in engineering and a Ph.D. did not apply in the electronics companies because there were very few engineers with this level of qualification.

In sum, qualification inflation within the occupation of technician had negative repercussions for technicians already employed within these firms. It served to harden the graduate barrier thereby condemning relatively young technicians to a career plateau (almost two-thirds of the technicians – 64.6 per cent – were less than thirty years of age). This restriction was resented by the technicians because it implied that their firm-related experience, and consequent ability to contribute to the organization's technical activities, were not as important as possession of a 'piece of paper'. Their employers no longer considered it worthwhile to extend the job hierarchy for technicians within the firm when more qualified labour was available on the external market. Not only was this considered unfair by the technicians, it was also a source of frustration because it implied that their careers had peaked. They now had to reconcile themselves to a future on the *technician plateau* even though they felt themselves to be quite capable of doing the work of engineers or chemists. This sense of frustration was made more acute by the experience of having expectations shattered shortly after taking up employment. The following advice was learned through bitter experience by the older technicians:

> 'If I met anyone now who is doing science I would tell them not to leave until they get a degree. You can still get hands-on experience with a degree but you are not given any kind of academic status [without one]. Years ago you had to have a Leaving Certificate [A-Level Equivalent], but now it is a case of a degree. You have to have one.'
>
> (Alice, technician, quality control, Harford Laboratories)

Given the various constraints of these branch plants, the technicians, engineers or chemists reached the limit of what was possible as a technical career within a number of years. Technical staff were faced with the option of becoming 'lifers', or leaving if they had no interest in moving into management. The more talented individuals could move elsewhere within the corporation provided they were prepared to emigrate. In this way these MNCs provided an international ILM that acted as a direct channel for the 'brain drain' that flows from countries on the economic periphery to those at the core (see Hanlon 1994 in the case of MNCs and Irish accountants). Since these were elite technical firms the career ladders are probably even shorter within the electronics sector generally. There is already evidence to indicate that the manufacturing operations of MNCs within the Irish electronics industry do not contain enough skilled engineering work to provide satisfactory career structures for experienced technical personnel (EOLAS 1989: 91). This then raises the question of the employers' approach to staff turnover.

The turnover 'problem'

Contrary to the emphasis on employee retention within the HRM and ILM literature, these employers held a remarkably positive view towards staff turnover. They were quite prepared to allow a not insignificant number of technical staff to leave for employment elsewhere. Overall, more than one-third (37.1 per cent) of the technical staff were either in the process of leaving their employer, or intended to do so in the near future. Kenine Pharmaceuticals had the lowest proportion with this response (22.2 per cent) while Trojan Electronics had the highest proportion of all (59.4 per cent). The exceptionally large response from the latter can be explained by the real fears held by staff about the future of the Irish plant as the corporation had experienced a dramatic fall in profitability over the preceding eighteen months.

The reasons for changing employers were those of typical 'cosmopolitans' (Gouldner 1957) who wished to advance their careers by moving from organization to organization in order to satisfy high-growth needs. Altogether, approximately one-third (35.8 per cent) of those who applied to another company stated that their main reason was career advancement while a similar proportion (37.8 per cent) indicated an interest in new work challenges (Table 7.2). The bulk of the Kenine Pharmaceuticals' respondents (75.0 per cent), who were mostly chemists and chemical engineers, cited career advancement. In each of the other firms a desire for new work challenges was the most popular or among the most popular responses. The single largest response in Trojan Electronics was, not surprisingly, 'job security' (36.5 per cent).

Contrary to the emphasis on retention in the managerial literature, managers in each of these firms emphasized the positive aspects

Table 7.2 Reasons for applying elsewhere

Company	*Kenine* %	*Harford* %	*Mertel* %	*Trojan* %	*Total* %
New work challenge	25.0	50.0	38.4	31.8	37.8
Career advancement	75.0	42.9	30.8	27.2	35.8
Job security	0.0	0.0	0.0	36.5	15.1
Higher pay	0.0	0.0	23.1	4.5	7.5
Location	0.0	7.1	7.7	0.0	3.8
Total	100.0	100.0	100.0	100.0	100.0
N=	4	14	13	22	53

of a steady turnover of technical staff. They also held an implicit notion of what constituted a suitable level of turnover, or equilibrium point, for their firm.

Managers in the chemistry-based firms, said that their problem with turnover was that, if anything, it was too low. In the electronics firms where there was a higher level of turnover, the view was that a regular 'bloodletting' – as distinct from a 'haemorrhage' – was beneficial to their stock of technical expertise. It allowed ageing knowledge and diminishing enthusiasm to be replaced with new ideas from eager graduate recruits. The differences in the level of turnover between the two industries were related to the range of technical activities that they carried out at branch plant level and to their pay policies. Both Kenine Pharmaceuticals and Harford Laboratories, as branch plants of major US corporations, offered particularly attractive employment opportunities for Irish graduates. As one manager explained:

> 'What keeps people here, I think for the really good performers it is the fact of the quality of the work that they are doing. I am under no illusion about it. . . . The pay here is extremely good. It tends not to be an issue simply because we pay well so nobody leaves here for higher salaries.'
>
> (chemistry section manager, technical operations, Kenine Pharmaceuticals)

The managers and employees in Harford Laboratories were conscious of the fact that they worked in what was frequently described as the 'largest private R&D unit in the country'. To change employers would inevitably mean a move away from R&D into more mundane chemistry work in other foreign-owned plants, or even worse in smaller Irish-owned firms. Both of these firms also had a reputation for being among the highest paying employers in the industry. This resulted from a policy of positioning their pay levels in the 'upper quartile' of the industry. This was done on the basis of information from annual industry salary surveys.

It was quite clear that these factors had resulted in a situation where the managers believed they were somewhat too successful in attracting and retaining technical staff. Both the personnel and the employee relations managers in Kenine Pharmaceuticals claimed that the low levels of turnover in the technical workforce had resulted in increasing levels of stagnation as the staff had become too familiar with the work of the plant as the years went by. The company had become a victim of its high salary position.

A similar view existed in Harford Laboratories, especially in the case of the technicians. A certain proportion of the chemists, chemical engineers and technicians were perceived to have lost the enthusiasm of their early years. They were subsequently no longer as productive as they had been in the past. The eagerness with which these graduates tackled their first job had diminished as the work eventually became more and more routine to them. Yet many were reluctant to leave for fresh challenges elsewhere as they had become quite comfortable with the lifestyles provided by their existing remuneration packages. Furthermore, a large proportion of them had begun to start families and had taken out mortgages. This combination made them economically, as well as professionally, dependent on these employers.

In contrast to the chemistry-based firms, both of the electronics firms had a policy of pitching their pay at the market average. Trojan Electronics had originally been among the highest paying employers in the electronics sector but the corporation's financial difficulties led it to adopt a more modest position. In Mertel Telecommunications it was widely accepted among the technical staff that the company 'doesn't hold onto the engineers'. A direct result of this outflow of engineers was that more than three quarters (79.3 per cent) of the technical personnel were less than thirty years of age. Similarly, in Trojan Electronics there was a perception that there was a high level of turnover among engineers, much higher than that of the technicians.

Support for the turnover in technical personnel was most clearly enunciated by managers in Trojan Electronics. They were quite explicit on the benefits of having a regular turnover of engineers and technicians. One senior manager put it this way:

> 'Turnover is good. It allows change to take place, it allows you to bring in new individuals and new blood into the organization and it allows promotional opportunities and things that go with that and that's good. If it is stagnant and there is no one leaving and no change, I think that is the worst situation you could have.'
>
> (business manager, Trojan Electronics)

This manager also explained that from the company's point of view it was important to recruit new graduates on an on-going basis so that it had access to the latest developments in electronic engineering. Given the rapid rate of technological change in microelectronics technology, they might even have experience of equipment that their predecessors did not know existed.

Contrary to the logic of recruit and retain it was the benefits of staff turnover rather than retention which were emphasized in these interviews. There were a number of reasons for this, all of which help to explain the tendency to recruit straight from college. First, it allowed employers to take in graduates who had more to prove than their older and more expensive counterparts. The more recent recruits were more willing to take on extra work and to give up their leisure time in order to complete tasks. Second, these new recruits would bring the latest technical developments with them. Third, they did not have to be retrained or 'reprogrammed'. Fourth, labour costs were reduced because most of the new appointments, who were recent graduates, entered at the bottom of the salary scales.

Furthermore, even though these individuals were knowledgeable in the technical activities of their employers, the managers in these firms did not believe that turnover damaged the body of expertise and tacit knowledge that had accumulated within the Irish plant. This was principally because the firm's pool of technical knowledge was not solely the possession of individual staff. The organization of work around project teams, the policy of having written reports on each project and the use of formalized standard operating procedures meant that the plant was able to hold enough technical information to ensure that no single employee was irreplaceable whatever their level of expertise.

Controlling turnover: selective retention

On the surface it might appear that these employers managed the pool of technical expertise in their organizations by simply exploiting the prevailing labour market conditions. The process was actually much more subtle. These firms did strive to retain staff but only those who were deemed to be particularly valuable to the firm because of their technical acumen and experience. Rather than viewing the over-arching strategy as one of crude labour market exploitation, it is best conceived as a mixture of market exploitation and selective retention.

Each of these organizations had specific practices for identifying and retaining those who were seen to be the key human resources. Two of the firms had even organized a committee of managers whose main functions were to identify key performers and develop policies to retain them. Three of the firms were also in the process of considering the extension of their existing technical career structures. This latter initiative was, however, dependent on the business strategy of the corporation and the role of the branch plants within that. Except for Harford Laboratories, these changes were restricted to the creation of a single additional position because of the ceiling on the complexity of the design and development work in the Irish plants. Where a formal extension of the existing career ladder was not possible, as in Kenine Pharmaceuticals, the problem of retaining key staff was overcome by providing salary adjustments and job rotation opportunities. These latter arrangements tended to be of an informal and personalized nature.

In Kenine Pharmaceuticals, the employee relations manager described management's ability to control the 'turnover valve' as they saw fit by holding onto some of those whom they considered most valuable while letting others go. This task was monitored by a special committee of senior managers, called the 'Human Resources Group', who met each quarter to formulate staff and succession planning policies. This group monitored the key performers within the Irish plant and devised career structures to retain them within the Irish operation or, failing that, within the corporation. Its chief purpose, according to the personnel manager, was to develop career routes for the more capable managerial and technical staff so as to combat the problem of limited career opportunities within the factory.

> 'We need to keep them moving because if you don't they're just going to get stale or else they're going to say "Well look there's no opportunities here for me and I'm going to leave and I'll go to Pfizer or I'll go somewhere else".'
>
> (personnel manager, Kenine Pharmaceuticals)

Part of this process was the implementation of a job rotation scheme that exchanged managers and some senior chemists between departments. The company had also made informal salary adjustments to accommodate certain senior chemists who wished to remain working in a technical capacity.

The informality of these arrangements was dictated by the lack of work considered appropriate for the introduction of a more senior technical position. Mertel Telecommunications, by contrast, was able to introduce such a position because of the arrival of more sophisticated project work in the Irish plant. Like Kenine Pharmaceuticals, it also had a human resource planning committee. The 'Key Resource Group' was established to identify the high fliers within the company and to map out possible career options for them. The reason for this, according to the personnel manager, was that engineers of that calibre were 'scarce objects'. When the research took place this firm was in the process of adding the position of 'principal engineer' onto the existing technical career path. The director of the technology unit admitted that the demands of the engineers were a factor in this decision. This was because their managers had become worried about the level of stagnation that was emerging within the group. Also, with the launch of the strategic technology group the position of a principal engineer could now be justified on the basis of work content. It was not, therefore, simply a case of the technology following individual career needs though employee pressures were important in bringing about this decision. Special salary arrangements were also made for other engineers who wished to remain within a technical role.

While management were, to some extent, influenced by the career needs of their staff their ultimate decisions were, as Causer and Jones (1993: 2) have also found, constrained by the business strategies of their corporations and, subsequently, by the work content available within the firm. This point was clearly illustrated in the case of Trojan Electronics. Its management had considered creating an additional position of 'principal engineer' partly in response to some demands from their engineers, but also because the Irish plant had gained a reputation for its engineering ability within the corporation. This reputation was a source of pride to the technical staff, not least because it was the reason why the Irish plant was saved from closure in 1989 while a sister plant in Scotland was shut. Even so, the senior managers decided, after some deliberation, that they could not introduce the position of principal engineer. The reason was that the future technical activities of the plant would not be sufficient to warrant the creation of an additional position. While the job content had been available in the past, the recent changes in the corporation's business strategy had reduced the need for much of the engineering work that had helped the Irish operation to build a strong reputation within the corporation. Trojan Electronics had decided to move away from a strategy of both designing and manufacturing computers. It had come to a realization that it could no longer afford to compete in the design, manufacture and sale of computers in the recession-bound electronics industry of the early 1990s. It would, instead, specialize in particular areas of manufacturing while subcontracting its PCB assembly work. Consequently, it would no longer require its previous levels of technical expertise.

Harford Laboratories, unlike the other three firms, did not face the same set of difficulties in meeting the career needs of key technical staff. While the others were able to satisfy these needs in the short term by introducing an additional position on the technical ladder, Harford Laboratories decided to introduce a technical career ladder. This contained a range of positions of comparable status to middle and senior management positions. They were in a position to do this because the Irish plant contained the corporation's European R&D headquarters. It had a strong R&D reputation within the corporation not least because it had, in the past, been responsible for the invention of some of the company's most successful commercial products. Unlike the other firms, Harford Laboratories' decision was influenced by the age of its research chemists. Many were approaching mid-career and had reached the upper salary limits of the existing job structures. This demographic reality and, more importantly, the fear of losing some of the more experienced research staff contributed to the company's decision to create a dual career ladder. The 'parallel path', which it introduced, consisted of a 'scientific path' with four positions arranged in a hierarchy to correspond with the status and salary scales of those on the 'managerial path'. The aim was to retain the more productive scientists without having them turn into 'modest' managers (Shepard 1958). In the words of the R&D Vice President: 'I think it was a wish on the part of the company to corner its best scientists because we need the creativity of good scientists. We don't need to convert a good scientist into a modest manager'.

Conclusion

This chapter set out to examine how the career needs of technical personnel were accommodated in the branch plants of multinational firms. An important feature of these firms was that the technical staff were still at an early stage in their careers. This meant that much of their efforts could be attributed to their desire to learn, to get project experience and to work their way up the organizational ladder.

Nevertheless, it would appear that these employers do not fit the assumptions of the literature on internal labour markets or the careful recruitment and retention model presented in the HRM literature on hi-tech firms. For example, in the case of scientists and engineers, human resource planning is perceived to be critical to employer competitiveness (e.g., Anderson and Kleingartner 1987; Katz 1988). This research has demonstrated that in favourable labour market conditions even elite employers of knowledge workers were prepared to exploit the benefits offered by a surplus of skill in the external labour market. Though there were differences in the work and in the product market strategies of chemistry and electronics-based employers, this did not lead to variations in their labour market strategies. The principal reason for this was the over-riding influence of the level of technical activity conducted within the branch plants and the prevailing labour market conditions. Their overall approach was shaped as much by traditional labour market considerations as by the value of individual technical expertise. These factors have arguably tended to be conflated in research that is conducted in labour markets characterized by skill shortages. Given the relatively weak labour market position of technical staff in Ireland, these employers had limited incentive to construct extended internal labour markets. The reason was that the entry and exit points between the internal and external labour markets were relatively fluid. More importantly, the flow itself was perceived to contribute to productivity in scientific–technical employment.

This was most clearly illustrated in the case of the technicians whose market situation was undermined from two directions. First, as I described in Chapter 3, the value of their 'technician level' qualifications held by the older technicians had declined as a result of qualification inflation within the national labour market. Second, while they once had the possibility of moving up through the firm's internal labour market this had become blocked by what Roberts *et al.* (1972: 215) call the graduate barrier (see also Whalley 1986a: 109; Crawford 1989: 138–145). Employers preferred to recruit more qualified and cheaper, but less experienced, engineers and chemists from the external labour market rather than promote internally. The primary reason for this was the comparatively higher wage costs associated with the technicians. This does not imply that the replacement costs involved were similar to those required for semi and unskilled staff; rather they were not prohibitive. Management were able to accept the loss of the tacit skills and expertise of individual staff in the knowledge that their existing work practices and routines would ensure that the general pool of technical expertise would not be seriously threatened.

The practices described here suggest an alternative labour market strategy to that found in HRM or ILM models of employment. This was a hybrid combination of labour market exploitation and selective retention that allowed employers to exploit the prevailing external and internal labour markets in order to sustain and regenerate their knowledge base. This strategy contained three key elements. First, a certain level of staff turnover was allowed, even desired, since it gave management the opportunity to continue to recruit what were perceived to be the best graduate engineers and chemists. Second, apart from Trojan Electronics, each of the firms had introduced arrangements to retain what were seen as key 'human resources' within the plant. In other words, career and remuneration packages were available for those employees who were considered too valuable to be allowed to leave. Some of the actual retention 'packages' were comparable to those identified by Zussman (1985: 141–142) and Whalley (1986a: 91) in that they were both informal and personalized. Third, these practices were primarily directed towards engineers, chemical engineers and chemists. There was less emphasis placed on retaining technicians in these firms because they were even easier to replace than chemists or engineers.

It should be noted that while this approach to the labour market in both industries was dependent on the career orientations of technical staff, it was made possible by the availability of technical graduates. Also, except for Harford Laboratories, most were aged between twenty-five and thirty-five years. They were still at a relatively early stage in their careers given that most of them had only emerged from higher education in their early twenties. They had not yet reached the age of forty which, according to Sofer's (1970: 273–274) research, is the point at which technical specialists become most concerned about their long-term career plans. If subsequent career demands should lead to increased labour turnover then these employers already have the labour market practices in place to deal with it. These practices may not only enable them to manage the turnover, they may also allow them to profit from it.

8

CONCLUSIONS

Introduction

Like other spheres of economic life, the world of work is characterized by constant change. The sociologist's task is to dig beneath the rhetoric, the buzzwords and the managerial acronyms in order to make sense of the nature and direction of these changes. In this study I set out to investigate whether certain contemporary management practices had transformed the traditional division between white-collar and blue-collar workers to a point where they share similar employment relations. More specifically, the aim was to assess whether the orthodox 'service contract' account of the employment relationship for professional and managerial employees is appropriate when the work situation of these employees has been subject to new forms of bureaucratization that seek to introduce market principles.

This particular argument has been made in a number of different studies. To recap, Savage *et al.* (1988; 1992) contend that changes in internal labour markets and the use of performance appraisals have significantly altered the traditional service relationship of managerial employees. Their claim that organizations have increasingly moved away from the use of internal hierarchies and towards various market mechanisms to structure their activities was of especial interest to this study. Similar claims have been advanced by Whittington (1990; 1991a; 1991b) in his study of R&D organizations. He claims that the 'externalization' of R&D work has reduced the autonomy previously enjoyed by professional scientists. Furthermore, Causer and Jones (1990; 1996) believe that the presence of project teams and performance-related pay mean that employers no longer simply trust professional white-collar employees. Smith's (1987) study of technical workers in British Aerospace also concludes that, apart from those at the very top of the technical hierarchy, the remainder could be classified as waged labour.

This study examined these arguments by adopting a *critical case* approach. This meant that the locales chosen for study were those which were most likely to validate the claims of change. The research was, accordingly, based on professional technical workers in bureaucratic employment since it is in such settings that these developments are most apparent. The selection of MNCs introduced a subsidiary question relating to the new international division of labour. The NIDL theorists claim that the type of manufacturing operations carried out by MNCs requires relatively limited skill levels. Previous research seemed to suggest that what technical activity existed in the branch plants of multinational firms was highly regulated and required relatively low skill levels. In other words, it is broadly supportive of the argument that employers do not simply trust professional and managerial staff to carry out their duties.

To assess these questions, this research focused on the policies and practices of employers and the experiences and attitudes of managers and their technical staff. These included management policies and practices for recruitment, work organization, supervision, remuneration and careers. Rather than simply repeat the findings presented in Chapters 3–7 I wish to turn to more general issues in this final chapter. The first part of the chapter assesses the impact of bureaucratization, in the form of HRM, on the service relationship. This assessment will be primarily concerned with professional technical employees (engineers and chemists). The distinctive situation of the technicians will be set out in the second part where I describe the implications of this research for the literature on HRM and for Ireland's experience of the new international division of labour.

Human resource management and 'service' employment

It was quite evident in this research that the traditional conception of service-class employment had changed as a result of employer practices. Two specific areas of managerial practice were at the source of these changes. These were the use of formal mechanisms for the coordination and monitoring of work and the changing balance between the internal and the external labour markets.

The impact of contemporary management practices

Each of the four MNC employers implemented a range of formal managerial policies that were designed to coordinate, monitor, evaluate and reward the work of their professional staff. In doing so, they extended the principles of bureaucratic rationalization into the employment of professional and managerial workers with the intention of achieving greater predictability, calculability and control in relation to their behaviour of these employees. While this is of interest in itself, it is of particular significance here because it shows that these employers did not simply trust their professional (and managerial) employees to act in their best interests.

Project management systems were one of the principal features of bureaucratic control for the professional staff (chemists and engineers). These were especially significant because they governed the overall direction of the work with the result that the engineers and chemists had little strategic autonomy i.e., what to work on (Bailyn 1985). Their work was entirely devoted to products and processes that had commercial applications. So, when compared to the firms studied by (Whalley 1986b) or (Whittington 1991a) it is not possible to speak of a shift towards market exposure since the hegemony of the market had long since been established in these plants. With respect to Mertel, Trojan and Kenine this emphasis was put in place when these organizations commenced operations in Ireland. In Harford, the longest established of these plants, this emphasis emerged only in the 1980s.

The work of all technical employees (professionals and technicians) was coordinated with this objective in mind. This was achieved through a mixture of modern bureaucratic managerial practices and traditional supervision. The engineers and chemists, and the technicians in two of the firms, had their work activities listed and recorded by the performance appraisal system. This, in theory, was subsequently used to monitor and evaluate their performance and, ultimately, to help make decisions over the allocation of performance-related pay and promotion. What is also interesting here is the extent to which these organizations tried to gather, record and use information on individual performance. The employee's file was then retrieved to inform decisions about promotion and, for the less fortunate, decisions about discipline or dismissal. An employee's behaviour thus became part of the organization's memory. When viewed in conjunction with the project management system, it would appear, as Savage *et al.* (1988) and

Causer and Jones (1990; 1996) have argued, that the traditional characterization of these employees as high-trust high-discretion is erroneous.

The staffing practices of these employers were also based on the principle of market exposure. Contrary to the 'best practice' HRM literature, these employers did not rely on their internal job hierarchies as a means of retaining and developing staff. Accordingly, the entry and exit points between the internal and external labour markets were quite fluid. This was most obvious in two areas. First, they recruited ever more qualified staff within specific technical occupations. Technicians were only appointed if they held NCEA diplomas while the engineers and chemists required postgraduate qualifications. This trend of qualification inflation placed a graduate barrier on the technicians' career path with the result that they could no longer move up the technical hierarchy into the position of engineer or chemist. Second, these employers chose to recruit cheaper and more qualified individuals from outside the firm rather than promote more experienced staff internally. While this was particularly true of the technicians, it was also the case for engineers and chemists who were frustrated by the lack of technical career opportunities within the branch plants. Apart from Harford, these plants were unable to provide the kind of advanced design and development activities that would warrant the creation of a technical hierarchy that had more than one step (i.e., from 'chemist' to 'senior' chemist, etc.). Consequently, some left to further their careers elsewhere. What was most significant here was that this was not a flow that these employers sought to stem, as might be expected from much of the existing literature on internal labour markets. Instead they exploited it as it enabled them to recruit 'fresh blood' cheaply while simultaneously avoiding the problem of holding onto employees who had become bored by the narrow range of technical activities undertaken in MNC branch plants.

This latter finding corresponds with the argument advanced by Savage *et al.* (1988) that employers of service class employees were increasingly relying on the external market rather than on hierarchical structures for managerial labour. In this case, it also includes professional labour. This development is important because it suggests that white-collar work, even of the professional kind, does not necessarily lead to a job for life. It is also interesting that these employers were not afraid of the costs incurred by the departure of such highly trained employees. The fact that no employee was irreplaceable can be attributed to bureaucratic management practices which retained much of the essential technical knowledge within the factory. The presence of such techniques as Good Manufacturing Practice, along with the requirements for TQM and Quality Assurance ensured that these organizations had formal standard operating procedures for a range of key technical activities. These practices, along with the lack of leading edge design and development work, made it relatively easy for these employers to absorb the costs associated with the loss of any one employee.

The service relationship revisited

It might be inferred from these observations that employers had, to paraphrase Perrow (1970), chosen to proliferate the work of their technical personnel with rules and reporting procedures rather than seeking to professionalize them. This was not the case. Instead these employers enjoyed the best of both worlds: they implemented a battery of management policies while simultaneously continuing to draw on the professional orientations of these employees. Despite the use of certain HRM practices, they still sought to retain elements of trust, discretion and status in critical areas of the white-collar employment relationship. This pattern is probably not unique to these branch plants. However, what the case studies indicate is that it was the most appropriate route given the nature of the work and, with that, management's inability to fully implement various policies.

The most distinctive feature of the work of these employees was its indeterminate and complex nature. Though it was only 'branch plant' technical work it still could not be organized around Tayloristic principles such as a detailed division of labour with highly simplified tasks. Even the work of the technicians, which was dominated by testing and repair routines, required knowledge of abstract principles and symbols and the ability to manipulate them. This mix of abstract knowledge and practical skill is, of course, one of the defining features of technician work (Barley and Orr 1997). Moreover, the electronics technicians also had to constantly update their knowledge so they could refine their methodologies for tracking down unusual problems or 'quirks' that new designs might produce. The work of the engineers and chemists was even more contingent and uncertain. This was not simply due to the lower levels of routinization and specialization when compared with the technicians. Much of their work drew on abstract, as opposed to practical, knowledge of electronics or chemistry. The analysis and interpretation of test results, the improvement of test procedures, the design of experiments and other such tasks required knowledge of the principles of their discipline. Since they were not driven by routines, they were responding constantly to new problems and situations. For example, the Irish plant may have been chosen as the first site to manufacture a new product within the corporation; existing products may have had to be reconfigured to meet the requirements of individual customers; or a customer installation (or problem) may have required an off-site trip to Germany, etc.

Consequently, it was not surprising to find that authority was expressed through a collegial style of management in which subordinates were treated as colleagues. With this style managers did not need to direct or monitor what employees were doing on a daily basis. In any case, it was difficult for them to know what their staff were working on due to the specialized nature of the work and the changing nature of the underlying technology itself. Offe's (1976) argument is relevant to this point. He claims that the internal complexity of organizations is such that there is a discontinuity of task between subordinate and superior. This makes it difficult for those in positions of authority to judge whether or not the goals have been achieved. In such situations, Offe (1976: 28) argues that management control is increasingly based on 'normative and ideological requirements' defined as necessary to the task but not related to its technical fulfilment.' In this case, the ideological element consisted of the professional orientations and values which the collegial approach controlled through a process of *management by expectation*. These expectations were based on shared values that were upheld by managers. These values were initially inculcated during university and higher education in a process of anticipatory socialization (Raelin 1985: 148–155). Raelin (1985) has described how student engineers, for example, visualize what life is like as an engineer while still at university. This activity, when combined with the experience of internship, acts to reinforce the development of professional values. The branch plant managers drew on these values and their own experience of technical work to set the standards. They then expected their staff to reach them and questioned their professional integrity if they did not. This form of control was so deeply embedded that colleagues effectively controlled each other. They all valued the respect of their peers and worked hard to achieve it. Finally, I would contend that this process operated at a deeper, and more powerful level, than the type of bureaucratic control exerted by corporate human resource policies because it shaped the daily informal social exchanges between manager and employee and between the employees themselves.

An important feature of this style of management was, as Fox (1974) predicted, a concern to accommodate the professional orientation of the engineers and chemists and, indeed, the technicians. There were two areas of managerial practice where this was most apparent. First, managers tried to ensure that, wherever possible, they allocated 'quality' work to staff, or else involved them in job rotation. Both practices sought to meet the motivational needs of these employees. Second, these employers tried to recruit employees who fitted into their criteria for acceptability. This included a strong personal work ethic, an ability to fit in with existing staff and a personality that suited the job on offer. In this way, they minimized the potential costs of depending (or trusting) on employees who could not be tightly controlled.

There is of course another reason why this pattern of employment relations persists. Despite the impressive developments in management policy, these are still incomplete as forms of labour control. The appraisal system contained a number of distinct weaknesses. The complexity and indeterminacy of the work were such that individual managers could not produce an objective evaluation of the work of their subordinates. Rapid technical changes within the industry also meant that these managers no longer understood what exactly their subordinates were doing. Furthermore, a problem with both the project and performance management systems was that they could become inflexible through over-planning. Rigid adherence would mean that employees were not free to move onto urgent problems that emerged in the production department or from customers. This did not happen. Nor was it desired. These policies were intended to act as an aid to managers. They probably never intended to take management out of the management process because flexibility among management and employees was simply too valuable in these highly competitive and complex industries. Finally, the proportion of the remuneration package that was performance-related was not so large as to indicate that there was significant individualization of reward. Instead it seemed that performance-related pay was important only at the margins. Given these limitations, these employers still had to trust their employees and provide them with considerable autonomy not only because it was necessary but also because it was more efficient. In short, the balance of the evidence confirms Goldthorpe's (1982: 168) original statement to the effect that the performance of these employees has less to do with 'the efficacy of external sanctions and rewards' and more with their degree of moral commitment.

Management scholars who are disciples of Foucault (1979) have, however, interpreted such measures as a distinctively modern form of disciplinary power that serves to observe, record and evaluate individuals in an exhaustive manner. Proponents of this view have drawn on workplace studies of total quality control (Sewell and Wilkinson 1992) and self-managing teams (Barker 1993) to show how employees have become the subjects of more sophisticated forms of surveillance. Yet it is not altogether clear that these studies of surveillance are really any different to those based upon the older concept of managerial control (Thompson and Ackroyd 1995: 626). If, however, these writers are given the benefit of the doubt then what does appear to be original is their claim that these forms of surveillance have simply turned the workplace into another 'total' institution (Goffman 1961). This is not surprising when it is remembered that Foucault (1979) developed his claims about disciplinary power through studies of prisons, asylums and hospitals.

The problem with this argument, however, is that it overestimates management's ability to manage. Management are assumed to be much more rational, knowledgeable and coordinated than is frequently the case. The image of management behaviour that emerges here is markedly different from that of the omniscient control-obsessed accounts presented by followers of Foucault or indeed those in the labour process tradition who were inspired by Braverman (1974). These findings complement those studies of managerial behaviour that have shown it to be less formal, rational and strategic than is often assumed (Mintzberg 1973; 1994; Watson 1994). Furthermore, this also applies to the use of those individual HRM techniques that might seem to be most representative of modern forms of bureaucratic surveillance. For instance, a study of managerial practice (see p. 20) concluded that managers were so concerned with their own short-term financial targets and objectives, particularly in downsized organizations, that they had neither the time nor the motivation to undertake proper appraisals of their subordinates (McGovern *et al.* 1998). What this means is that it would be naive to interpret the presence of particular HRM policies in the white-collar work situation as evidence of substantial change in service-class employment. The mere introduction of appraisals and performance-related pay does not somehow transform the underlying pattern of social relations in the workplace. Even if every manager conducted their appraisals in the manner set out by the company handbooks, their inability to specify precise work objectives in advance meant that managers still had to manage in a collegial rather than autocratic style.

The argument that I have made so far has emphasized the need for a more realistic view of the nature of managerial practice which is based on empirical evidence and not simply on fashionable theories. There is, of course, an element which relates to employee responses to such practices. This was not always one of willing acceptance. Instances of collective and individual resistance appeared at different points. While there is, according to Thompson and Ackroyd (1995), a tendency to ignore the role played by employee resistance in Foucauldian studies of new management practices, this can only lead to a blinkered perspective. Employee resistance within the capitalist enterprise, as Paul Edwards (1986; 1992) and others have argued, is nearly always present in one form or another. Technical work is no exception.

First, the technicians' union in Harford Laboratories and Kenine Pharmaceuticals prevented the introduction of performance appraisals and performance-related pay. Although their employers would have liked to extend these policies to the technicians, the technicians refused to consider them because they were incompatible with the collectivist orientations of their trade unions (MSF and SIPTU respectively). Their view was that pay increases should be awarded on a collective rather than on an individual basis. They also feared that the individualized systems would be open to abuse by managers who disliked certain technicians. Second, complaints by individual chemists and engineers in Kenine and Harford had forced these employers to address the problem of status inconsistencies in pay. If they had not done so, it is questionable whether these employers would have initiated the 'knock-ons' that followed from the wage increases won by the technicians' unions. Third, the bureaucratic process that led to 'the numbers game' in Trojan Electronics was initially disputed by the technicians because it undermined the technical content of their role. Finally, it should also be remembered that these employers were reluctant to break with the established link between occupations and qualifications in recruiting new staff because of the fear that it would be resisted by employees already within the organization.

These various forms of employee resistance reinforce the argument that the presence of new management practices in a work situation does not necessarily lead to fundamental change. Similarly, processes of bureaucratization are not always an inevitable phenomenon in every circumstance. Their implementation may be prevented, curtailed or treated cynically even when in place. For instance, one of the technicians in Mertel Telecommunications viewed the Managing For Achievement appraisal system as simply another piece of meaningless 'bureaucracy' (see p. 87 for his alternative interpretation of MFA).

Much of this research, along with that of others, has focused on the nature of the changes in the white-collar service contract that have resulted from contemporary management practices. Given this emphasis, there is a danger that those practices that actually maintained traditional elements of the employment relationship may be overlooked. This would be misleading, especially when these employers consciously tried to preserve existing status differences in their employment practices. Despite the favourable labour market conditions, these MNCs did not seek to recruit those who had trained for the position of engineer or chemist into that of technician. To do so would have broken with the prevailing labour market conception of occupational status. They also tried to ensure that their pay practices reflected this difference in employment status. When the pay bargaining of the technicians' unions led to status inconsistencies in Harford and Kenine, these employers took measures to restore the original differentials between technicians and chemists. Moreover, these employers offered the prospect of a career, even if it was one that contained a low ceiling. The possibility of a career, as distinct from a job, has traditionally been one of the key differences between white and blue-collar jobs. There was no evidence in this research that this had changed.

It would be wrong, however, to assume that this research is completely supportive of the existing theoretical treatment of the service relationship. There are two minor qualifications to be noted. First, it is typically assumed that those in service-class employment have greater job security than those employed on waged labour contracts (Goldthorpe 1982). According to Breen (1997) firms typically enter into long-term commitments with these employees. However, this assumption pre-dates the arrival of delayering as a form of corporate downsizing. Delayering is, according to Heckscher (1995), a deliberate assault on white-collar employees because they represent the 'corporate fat'. Delayering rests on the premise that once this fat is trimmed, corporations become leaner, meaner and more competitive. While the validity of this employer strategy may be challenged it has marked a watershed: it is the job security of service class employees that has been undermined and not that of waged labour. It was professional and managerial staff who were made redundant in Harford and Trojan and not the semi-skilled operatives or clerical staff. Having said that, there is a strong possibility that 'delayering' is a phenomenon that is unique to large corporations (and possibly only those who believe the latest nostrums from management gurus and consultants).

What is more significant is that a 'service relationship' can exist in situations where employers are not concerned with retaining highly skilled employees. The knowledge of these organizational professionals was not considered to be sufficiently specific, and consequently of such high value, to these employers that they made efforts to retain them. This can probably be attributed to labour market conditions that clearly suited these firms. But despite these conditions, these employers still felt it necessary to maintain the 'service relationship'. Though some of their practices may have stemmed from tradition or from within the corporation, this approach was made necessary because of the nature of the work and by the limitations of their own policies.

The second qualification relates to the idea that the service relationship should be understood as the means through which an employer seeks to create and sustain commitment (Erikson and Goldthorpe 1993: 42). The conditions of white-collar employment (salary, promotion prospects) may help to create and sustain such commitment but they are not sufficient. Other factors relating to the work situation such as employer practices and the organization of work are also critical. In Chapter 4 I found that such commitment cannot be assumed where the actual work is repetitive, tedious and lacking in challenge. 'Quality work' was of particular importance in motivating these workers. If the work was interesting, if it was providing an opportunity for learning new knowledge then it helped to create and sustain commitment.

My overall argument, in summary, is that the introduction of new management practices had indeed changed professional employment. Employers no longer simply trusted employees to execute their responsibilities because they had implemented policies to formally monitor, evaluate and reward their work. While there have indeed been changes I have repeatedly emphasized the point that the apparent bureaucratization has not transformed the traditional underlying pattern of employment relations. This was because of the limits of managerial practice and, in some areas, because of actual employee resistance to new practices. So while there have been some changes their impact on the service relationship has clearly been greatly exaggerated.

HRM, multinational corporations and the 'Celtic Tiger'

The primary focus of this research has been on the impact of contemporary managerial practices on white-collar employment. However, I also indicated at the outset that this research would seek to contribute, in a more modest way, to our understanding of HRM and of the role of MNCs in late developing countries.

The costs of commitment

The theme of employee commitment has been one of the central strands of human resource management since it emerged from the US in the 1980s (Guest 1987; Storey 1992; Legge 1995). The message is that organizations can procure commercial success by creating and maintaining a high commitment workforce. The evidence produced in this research challenges some of the assumptions within this prescription. First of all, the concept of high commitment is presented in an unproblematic fashion. High commitment is not always a management panacea because managers may actually encounter problems in controlling commitment. More specifically, evidence was presented which showed that professional white-collar employees present their employers with problems precisely because of their high levels of commitment. For instance, they became demotivated when they did not get work that met with their professional interest in, and personal enthusiasm for, electronics or chemistry. Efforts to raise commitment solely through HRM policies (training, pay, development, etc.) will have a limited impact on the performance of knowledge workers if they fail to focus on the intrinsic aspects of their employment, especially in the areas of job design and development (see also Morris *et al.* 1993). As Chapter 6 indicated, these workers also became disgruntled when their pay did not match their employment status. Some were also more than happy to leave if they could advance their careers elsewhere. Then of course there is the possibility of conflict between professional values and organizational requirements. This is an old theme in the literature on organizational professionals (e.g., Kornhauser 1962). It was most vividly illustrated in this research by the conflict between the technicians and their managers over the 'numbers game' in Trojan Electronics.

Yet none of these problems have been recognized within the HRM literature. It could be argued here that these problems are peculiar to highly educated employees who have to work in MNC branch plants. To do so, however, would be to avoid the point. High levels of commitment may lead to competing forms of commitment, especially when such commitment is derived from professional or occupational sources. Guest (1987) appears to be alone in acknowledging this possibility within the UK HRM literature. He has, for instance, also examined this issue in a study of dual commitment (Guest and Dewe 1991). Even where this commitment is not based on allegiance to another reference group, employers still have to satisfy the needs of individual employees. Those who invest heavily of themselves in an organization may expect this to be returned. Commitment is, in other words, a double-edged sword.

The employer may not, however, be as willing to reciprocate this level of commitment if alternatives are available. An example of this was the attitude that the four MNCs adopted towards employee turnover. Where best practice models of HRM would emphasize careful recruitment and retention, these employers practised careful recruitment and replacement. They did so because labour market conditions were such that it was frequently cheaper to recruit externally rather than always promote internally. What this suggests is that best practice HRM models may not meet the needs of employers who wish to have high quality labour but at the lowest possible cost.

This pattern of behaviour is best understood as the result of a structured antagonism between labour as a commodity, and capital as a purchaser of that commodity (Edwards 1986: 5). While work relations may be characterized by cooperation and commitment, it also includes the potential for conflict. HRM like any other form of labour management also has to deal with these tensions (Legge 1989: 38). Its capacity to empower employees will only be partially successful where the employment relationship is founded on different interests between capital and labour. This structured antagonism is important here in that the utilization of technical labour for the purpose of creating profit may limit the extent to which employers are able, or wish, to accommodate work and career needs. The business environment, the structure of the firm and the organization of technical work within it were such that the expectations and aspirations of technical staff could be only partially satisfied.

Divisions of labour

The selection of MNCs raised issues connected to theories of a new international division of labour. To reiterate, this perspective claims that MNC branch plants are attracted to the peripheral economies of the global economic system because they offer a plentiful supply of cheap labour (Frobel *et al.* 1980; Henderson 1989). Once there, they proceed to establish mass production factories whose manufacturing processes require little more than semi-skilled labour. This line of argument has, as indicated in Chapter 1, become well established in studies of foreign direct investment in the Irish Republic (e.g., Perrons 1981; 1986; Sklair 1988). There is, of course, an alternative view promoted by Irish development agencies, politicians and others with a vested interest which claims that Ireland is rapidly becoming the hi-tech centre for Europe. It has achieved this by offering an abundant supply of highly educated labour and incredibly low levels of corporate tax, a combination that has attracted virtually all the leading American corporations. Since these global firms are all at the forefront of R&D they employ large proportions of university graduates and skilled workers in their Irish operations.

On the basis of this research, the reality lies closer to the NIDL account. The firms examined here represented the technical elite of two industrial sectors yet only one of these contained a genuine R&D unit. Only Harford undertook the kind of research that would lead to the development of new products (and patents) in Ireland. In the others, much of the leading edge technical activity consisted of modifications to existing products and processes. These were generally demanded by customers who, for example, wanted North American products reconfigured to meet Scandinavian standards. While neither Kenine nor Mertel contained R&D facilities, both had developed from simply providing technical support to the production function to a situation where they engaged in some development work for the Irish plant and also for sister plants in the European market. Trojan Electronics, by contrast, was strictly an assembly operation in which the technicians and engineers were deployed either directly in production or in supporting roles. Their lack of design and development work was reflected in the experience of work. Those in the two electronics firms, in particular, were frustrated by a lack of interesting technical work. The production technicians were affected the worst as their work consisted of repetitive testing and repair procedures. Their managers tried to deal with this problem as best they could through job rotation. But even this was constrained by the work on offer. It follows that if this is the situation in those firms that constitute the technical elite then the work must be of an even more limited nature in the general population of MNC plants.

It should not be inferred from this that these plants will only ever be 'screwdriver' operations. Harford Laboratories did not have an R&D facility when it was originally established in Ireland. It was however some ten years after it commenced production before the R&D unit was launched. Kenine Pharmaceuticals and Mertel Telecommunications had, as indicated in Chapter 4, also increased the depth of their design and development activities. By contrast, the future in Trojan Electronics looked particularly bleak as changes in business strategy meant that it would no longer require the same level of technical personnel (the plant was closed eighteen months after this research was completed). These developments were, apart from Trojan, warmly welcomed by those employed in the Irish plants because it meant that they have the opportunity of more interesting work. What was even more important was that these changes indicated that the Irish site was becoming more important within the corporation. It would therefore be less likely to be closed should there be a significant dip in profits.

Yet there was no evidence to suggest that significant amounts of R&D work had been transferred from corporate centres to the Irish sites. However, the changes that did occur related to (production) process development rather than product research and design activities. These changes are unlikely to provide years of exciting new technical activity for a remarkably young workforce. Nor are they likely to result in significant extensions to the technical career ladders within these firms. In these circumstances, it would not be a surprise if these particular members of the 'Celtic Tiger' were to become increasingly bored and frustrated by the lack of career progress and the lack of leading edge technical activity as they approach middle age.

Finally, it was also clear that the employment structures and practices of these MNCs reflected more general changes taking place in the class structure of Irish society. According to Breen *et al.* (1990: 138–140) the growing specialization and differentiation of the labour force in tandem with the expansion of the Irish education system has led to the increased use of educational qualifications in the selection of individuals for jobs. This trend is arguably most clearly manifested in the MNC-dominated hi-tech sector of the Irish economy. This development can be traced through the way jobs are allocated in hi-tech MNC plants and in the subsequent treatment of these employees under a dualistic system of employee relations. For entry into jobs the principal sorting device is the Irish education system. Educational qualifications determined who fitted into what position. Those with secondary school qualifications were employed either as semi-skilled production workers, craft workers or in secretarial and clerical positions. Those with diplomas and certificates from higher education institutions were employed as technicians while those who held university degrees and higher worked as engineers or chemists and managers. This sorting device raised invisible walls within these organizations. Shopfloor employees could not hope to move into salaried positions. Technicians were unlikely to ever become an engineer or chemist whatever their practical talents. The rapid credentialling of the technical labour market has ensured that employees can achieve little in the 1990s without the 'piece of paper'. While employees may be attracted to these organizations because they provide an attractive range of career opportunities, the reality, in the words of Randall Collins (1979: 43) 'is closer to a castelike separation among major occupational blocs'.

These divisions were reinforced by a dualistic system of labour management. Those with third-level qualifications were paid salaries, had interesting jobs, could expect careers and were not subject to direct supervision. Those who worked in production were paid wages, had routine jobs and had to obey the directions of their supervisors. These employees, apart from those in Trojan Electronics, consequently had to advance their interests through trade unions.

The slight exception to this was the technicians. They occupied an 'intermediate position' although this was marked by a divergence between their middle-class market and work situations (see also Brown 1995). They enjoyed considerable labour market power by virtue of their possession of third-level educational qualifications; they had the prospect of a career rather than a job; and they were also motivated by the opportunity of doing 'quality work'. Yet they were paid wages, 'clocked in', belonged to trade unions (Harford and Kenine) and had their work, which was dominated by routines, directed by supervisors. Furthermore, they were unlikely to hold managerial responsibilities and, in that sense, were less trusted by management (see also Creighton and Hodson 1997). In these respects they are best viewed as 'technical labourers' rather than as 'elite human resources' (Jones *et al.* 1993) when compared with the engineers and chemists. They were also the victims of the increase in qualification standards. Prior to the mid-1980s, they could have become associate engineers and chemists and possibly even managers. Though they still had the prospect of a career it was one that would only take them to the top of a ladder that contained only two rungs.

Such details tend to be forgotten in the rush by successive governments and various state agencies to show MNCs just how hospitable the Irish are. Potential MNC investors are offered capital grants, R&D grants, training grants, huge tax concessions and an abundant supply of young, highly educated, English speaking employees. Such hospitality is not only expensive, it is also extremely naive. Perhaps that is why politicians and development officials find it necessary to promote the myth that Ireland attracts hi-tech firms who employ significant numbers of graduates in their R&D facilities. The four firms that participated in this research were selected because they represented the technical elite. Yet only one of these actually undertook something approaching genuine R&D activity. Another (Trojan Electronics) was subsequently closed even though the Irish plant had been quite profitable. In the words of a manager in Mertel Telecommunications, 'there is no genuine R&D in Ireland'. It is unlikely that there ever will be much so long as the best engineers, chemists and technicians are employed by the branch plants of multinational corporations.

APPENDIX

Semi-structured interviews

From the companies' viewpoint, the semi-structured interviews were a rather expensive exercise because they had to 'write-off' the time that their technical staff spent being interviewed. Some were initially reluctant to participate in the research for this reason. Fortunately, they eventually agreed to become involved without charging the labour costs to my (insignificant) fieldwork budget! The interviews lasted between forty minutes and one and a half hours. Different sets of questions were used for managers, employees, shop stewards and members of the personnel departments. Those with managers tended to be longer as they provided detailed information on various policies and practices. All but one of the interviews were recorded. A total of seventy-nine people were interviewed in the first phase of the research which lasted from November 1990 to June 1991. The following is a list of interviews by firm, position and department:

Kenine Pharmaceuticals, 22

Position: 9 managers (1 × 2), 1 supervisor, 4 chemists, 4 chemical engineers, 2 technicians, 1 general operative (Chairman SIPTU committee).

Department: 5 personnel management, 5 quality control, 8 technical operations, 2 planning, 2 trade union representatives.

Harford Laboratories, 16

Position: 6 managers, 2 chemists, 1 chemical engineer, 2 technical specialist, 5 technicians.

Department: 3 personnel management, 3 quality control, 8 R&D, 2 trade union representatives.

Mertel Telecommunications, 22

Position: 6 managers (1 × 2), 9 engineers, 4 technical specialists, 2 technicians.

Department: 2 personnel management, 5 quality control, 10 technology unit, 3 manufacturing engineering, 2 manufacturing.

Trojan Electronics, 20

Position: 5 managers, 2 supervisors, 5 engineers, 1 associate engineer, 7 technicians.

Department: 2 personnel management, 4 quality control, 10 manufacturing, 4 manufacturing engineering.

Analysis of semi-structured interviews

All seventy-nine interviews were transcribed over a period of eleven months. The interview material was then formatted, sorted and categorized using the ETHNOGRAPH software package for the analysis of text-based data. Though this procedure took some time, the software ultimately proved to be much faster and more economical than the old-fashioned scissors, glue and index cards. The disadvantage was that my use of ETHNOGRAPH's search facility produced huge piles of transcript with excerpts from numerous different interviews.

The coding categories were derived from the subject of the question (key words) and from the answers, especially where they varied. All of the interviews were then searched under various codes to provide information on specific issues. This frequently produced 10–12 similar sets of comments from different individuals on the same topic. I have only included 2–3 quotes on any specific topic for reasons of space. Consequently, I have only quoted excerpts from a selection of the seventy-nine interviews.

Questionnaire survey

The questionnaire schedule was developed using interviews from the first stage of the research. It contained sixty questions divided into nine sections: Personal Details; Educational Qualifications and Work History; Recruitment and Training; Attachment to Employer; Technical Work and Responsibilities; Discretion and Supervision in Technical Work; Technical Standards and Performance Levels; Personnel Policies and Technical Work; Professional and Technical

Identity. The questionnaire was self-administered and took approximately twenty-five minutes to complete.

Some 203 questionnaires were distributed in person to technical employees in the four firms (total population of 257). The data was coded and entered using the PC DATA ENTRY programme and the results were analyzed using SPSS. I have only included those items (contingency tables) that were most relevant to the focus of this study. Details of these and other questionnaire items along with the questions used during the semi-structured interviews can be obtained by writing to the author.

BIBLIOGRAPHY

AGCSI (1989) *Graduates and their Careers*, Dublin: Association of Graduate Careers Services in Ireland.

Althauser, R. P. and Kalleberg, A. L. (1981) 'Firms, Occupations, and the Structure of Labor Markets: A Conceptual Analysis', in I. Berg (ed.) *Sociological Perspectives on Labor Markets*, New York: Academic Press.

Anderson, C. S. and Kleingartner, A. (eds) (1987) *Human Resource Management in High Technology Firms,* Lexington: Lexington Books.

Armstrong, P. (1987) 'Engineers, Management and Trust', *Work, Employment and Society* 1, 4: 421–440.

—— (1989) 'Management, Labour Process and Agency', *Work, Employment and Society* 3, 3: 307–322.

—— (1995) 'Accountancy and HRM', in J. Storey (ed.) *Human Resource Management: A Critical Text*, London: Routledge.

Armstrong, P., Carter, R., Smith, C. and Nichols, T. (1986) *White Collar Workers: Trade Unions and Class*, London: Croom Helm.

Bailyn, L. (1985) 'Autonomy in the Industrial R&D Lab', *Human Resource Management* 24, 2: 129–146.

Bailyn, L. and Lynch, J. T. (1983) 'Engineering as a Life-long Career: Its Meaning, Its Satisfactions, Its Difficulties', *Journal of Occupational Behaviour* 4, 4: 263–283.

Baldamus, W. (1961) *Efficiency and Effort: An Analysis of Industrial Administration*, London: Tavistock.

Barker, J. R. (1993) 'Tightening the Iron Cage: Concertive Control in Self-Managing Teams', *Administrative Science Quarterly* 38: 408–437.

Barley, S. R. and Orr, J. E. (eds) (1997) *Between Craft and Science: Technical Work in US Settings*, Ithaca: ILR Press.

Beer, M., Spector, B., Lawrence, P. R., Quinn Mills, D. and Walton, R. E. (1985) *Human Resource Management: A General Manager's Perspective*, Glencoe, Ill.: Free Press.

Blackburn, R. and Mann, M. (1979) *The Working Class in the Labour Market*, Basingstoke: Macmillan.

Blauner, R. (1964) *Alienation and Freedom*, Chicago: University of Chicago Press.

Blyton, P. and Turnbull, P. (eds) (1992) *Reassessing Human Resource Management*, London: Sage Publications.

Bosworth, D., Wilson, R. and Taylor, P. (1992) *Technological Change: The Role of Scientists and Engineers*, Aldershot: Avebury.

Braverman, H. (1974) *Labour and Monopoly Capital*, New York: Monthly Review Press.

Breen, R. (1997) 'Risk, Recommodification and Stratification', *Sociology* 31, 3: 473–489.

Breen, R., Hannan, D. F., Rottman, D. B. and Whelan, C. T. (1990) *Understanding Contemporary Ireland: State, Class and Development in the Irish Republic*, London: Macmillan.

Brown, P. (1995) 'Cultural Capital and Social Exclusion: Some Observations on Recent Trends in Education, Employment and the Labour Market', *Work, Employment and Society* 9, 1: 29–52.

Brown, P. and Scase, R. (1994) *Higher Education and Corporate Realities: Class, Culture and the Decline of Careers*, London: UCL Press.

Buckley, P. and Enderwick, P. (1985) *The Industrial Relations Practices of Foreign-Owned Firms in Britain*, London: Macmillan.

Burawoy, M. and Wright, E. O. (1990) 'Coercion and Consent in Contested Exchange', *Politics and Society* 18, 2: 251–266.

Burns, T. and Stalker, G. M. (1961) *The Management of Innovation*, London: Tavistock.

Butler, T. and Savage, M. (eds) (1995) *Social Change and the Middle Classes*, London: UCL Press.

Causer, G. and Jones, C. (1990) 'Technical Workers, Work Organization and Career Structures in the Electronics Industry', Paper presented to the *Organization and Control of the Labour Process Conference*, Aston University, 28–30 March.

—— (1993) 'Responding to "Skill Shortages": Recruitment and Retention in a High Technology Labour Market', *Human Resource Management Journal* 3, 3: 1–20.

—— (1996) 'Management and the Control of Technical Labour', *Work, Employment and Society* 10, 1: 105–123.

Child, J. (1981) 'Professionals in the Corporate World: Values, Interests and Control', in D. Dunkerley and G. Salaman (eds) *International Yearbook of Organizational Studies*, London: Routledge.

Clark, J., McLoughlin, I., Rose, H. and King, R. (1988) *The Process of Technological Change*, Cambridge: Cambridge University Press.

Cogan, D. J. and O'Shea, F. (1979) *Engineering Manpower Survey*, Dublin: Science Policy Research Centre, University College, Dublin.

Collins, R. (1979) *The Credential Society: An Historical Sociology of Education and Stratification*, New York: Academic Press.

Cotgrove, S. and Box, S. (1970) *Scientists and Industry*, London: Allen and Unwin.

Crawford, S. (1989) *Technical Workers in an Advanced Society: The Work, Careers and Politics of French Engineers*, Cambridge: Cambridge University Press.

Creighton, S. and Hodson, R. (1997) 'Whose Side Are They On? Technical Workers and Management Ideology', in S. Barley and J. E. Orr (eds) *Between Craft and Science: Technical Work in US Settings*, Ithaca: ILR Press.

Crompton, R. (1980) 'Class Mobility in Modern Britain', *Sociology* 14, 1: 117–119.

Crompton, R. and Jones, G. (1984) *White-collar Proletariat: Deskilling and Gender in Clerical Work*, London: Macmillan.

Crompton, R. and Jones, G. (1988) 'Researching White-collar Organizations: Why Sociologists Should Not Stop Doing Case Studies', in A. Bryman (ed.) *Doing Research in Organizations*, London: Routledge.

Devanna, M. A., Fombrun, C. and Tichy, N. M. (1984) 'A Framework for Strategic Human Resource Management', in C. Fombrun, N. M. Tichy and M. A. Devanna (eds) *Strategic Human Resource Management*, New York: Wiley.

Doeringer, P. B. and Piore, M. J. (1971) *Internal Labor Markets and Manpower Analysis*, Lexington: Lexington Books.

Dore, R. (1976) *The Diploma Disease: Education, Qualification and Development*, London: Allen and Unwin.

Economist (The) (17.5.1997) 'Europe's Tiger Economy', 25–28.

Edwards, P.K. (1986) *Conflict at Work: A Materialist Analysis of Workplace Relations*, Oxford: Basil Blackwell.

—— (1990) 'Class and Work: In Search of the Connections', *Warwick Papers in Industrial Relations*, No. 17, IRRU, School of Industrial and Business Studies, University of Warwick.

—— (1992) 'Industrial Conflict: Themes and Issues in Recent Research', *British Journal of Industrial Relations*, 30, 3: 361–404.

Edwards, R. (1979) *Contested Terrain: The Transformation of the Workplace in the Twentieth Century*, London: Heinemann.

Eisenhardt, K. M. (1989) 'Building Theories from Case Study Research', *Academy of Management Review* 14, 4: 532–550.

—— (1991) 'The Case for Rigor and Comparative Logic', *Academy of Management Review* 16, 3: 620–627.

EOLAS (1989) *Electronics Manpower Survey: Trends in the Irish Electronics Manufacturing Industry up to 1995*, Dublin: EOLAS.

Erikson, R. and Goldthorpe, J. H. (1993) *The Constant Flux: A Study of Class Mobility in Industrial Societies*, Oxford: Oxford University Press.

Etzioni, A. (1961) *A Comparative Analysis of Complex Organizations*, New York: The Free Press.

Evans, G. and Mills, C. (1997) 'In Search of the Wage-Labour/Service Contract', Paper presented to the *ISA Research Committee on Social Stratification*, Tel Aviv, May.

—— (1998) 'Identifying Class Structures: A Latent Class Analysis of the Criterion-Related and Construct Validity of the Goldthorpe Class Schema', *European Sociological Review* 14.

Fielding, N. G. and Lee, R. M. (1991) *Using Computers in Qualitative Research*, London: Sage.

Flood, P. and Toner, B. (1997) 'Large Non-Union Companies: How Do They Avoid a Catch 22?', *British Journal of Industrial Relations* 35, 2: 257–277.

Flood, P., Toner, B. and Turner, T. (1994) 'Human Resource Strategy and the Non-Union Phenomenon', in P. Gunnigle, P. Flood, M. Morley and T. Turner (eds) *Continuity and Change in Irish Employee Relations*, Dublin: Oak Tree Press.

Foucault, M. (1979) *Discipline and Punish: The Birth of the Prison*, Harmondsworth: Penguin.

Fox, A. (1974) *Beyond Contract: Work, Power and Trust Relations*, London: Faber and Faber.

Friedman, A. (1977) *Industry and Labour: Class Struggle at Work and Monopoly Capitalism*, London: Macmillan.

Frobel, F., Heinrichs, J. and Kreye, O. (1980) *The New International Division of Labour*, Cambridge: Cambridge University Press.

Gallie, D. (1978) *In Search of the New Working Class*, Cambridge: Cambridge University Press.

Geary, J. F. (1991) 'Human Resource Management in Practice: Labour Management in Irish Electronics Plants', unpublished D.Phil. thesis, University of Oxford.

—— (1992) 'Pay, Control and Commitment: Linking Appraisal and Reward', *Human Resource Management Journal* 2, 4: 36–54.

Giddens, A. (1973) *The Class Structure of the Advanced Societies*, London: Hutchinson.

Goffman, E. (1961) *Asylums*, Harmondsworth: Penguin.

Goldthorpe, J. (1980) *Social Mobility and Class Structure in Modern Britain*, Oxford: Oxford University Press.

—— (1982) 'On the Service Class, its Formation and Future', in A. Giddens and G. MacKenzie (eds) *Social Class and the Division of Labour*, Cambridge: Cambridge University Press.

—— (1995) 'The Service Class Revisited', in T. Butler and M. Savage (eds) *Social Change and the Middle Classes*, London: UCL Press.

Goldthorpe, J. and Whelan, C. (eds) (1992) *The Development of Industrial Society in Ireland*, Oxford: Oxford University Press.

Goldthorpe, J. H., Lockwood, D. and Platt, J. (1968) *The Affluent Worker: Industrial Attitudes and Behaviour*, Cambridge: Cambridge University Press.

Gomez-Mejia, L. R. and Balkin, D. B. (1989) 'Effectiveness of Individual and Aggregate Compensation Strategies', *Industrial Relations* 28, 3: 431–445.

Gouldner, A. (1957) 'Cosmopolitans and Locals: Towards an Analysis of Latent Social Roles', *Administrative Science Quarterly* 2: 281–306.

Grint, K. (1993) 'What's Wrong with Performance Appraisals? A Critique and a Suggestion', *Human Resource Management Journal* 3, 3: 61–77.

Guest, D. (1987) 'Human Resource Management and Industrial Relations', *Journal of Management Studies* 24, 5: 503–521.

Guest, D. and Dewe, P. (1991) 'Company or Trade Union: Which Wins Workers' Allegiance?', *British Journal of Industrial Relations* 29, 1: 75–96.

Gunnigle, P. (1992) 'Management Approaches to Employee Relations in Greenfield Sites', *Irish Business and Administrative Research* 13: 44–58.

Gunnigle, P., Morley, M. and Turner, T. (1997) 'Challenging Collectivist Traditions: Individualism and the Management of Industrial Relations in Greenfield Sites', *Economic and Social Review* 28, 2: 105–134.

Gunnigle, P., Flood, P., Morley, M. and Turner, T. (1994) *Continuity and Change in Irish Employee Relations*, Dublin: Oak Tree Press.

Gunz, H. (1980) 'Dual Ladders in Research: A Paradoxical Organizational Fix', *R&D Management* 10, 3: 113–118.

Halford, S. and Savage, M. (1995) 'The Bureaucratic Career: Demise or Adaptation?', in T. Butler and M. Savage (eds) *Social Change and the Middle Classes*, London: UCL Press.

Handy, C. (1989) *The Age of Unreason*, London: Business Books.

Hanlon, G. (1994) *The Commercialisation of Accountancy: Flexible Accumulation and the Transformation of the Service Class*, Basingstoke: Macmillan.

HEA (1983–1991) *First Destination of Award Recipients in Higher Education*, Dublin: Higher Education Authority.

Heckscher, C. (1995) *White-collar Blues: Management Loyalties in an Age of Corporate Restructuring*, New York: Basic Books.

Heisig, U. and Littek, W. (1995) 'Trust as a Basis of Work Organization', in W. Littek and T. Charles (eds) *The New Division of Labour*, New York: de Gruyter.

Henderson, J. (1989) *The Globalization of High Technology Production*, London: Routledge.

Hill, S. (1991) 'Why Quality Circles Failed but Total Quality Management Might Succeed', *British Journal of Industrial Relations* 24, 4: 541–568.

Hirsch, P. (1993) 'Undoing the Managerial Revolution?', in R. Swedberg (ed.) *Explorations in Economic Sociology*, New York: Russell Sage Foundation.

Hodson, R. and Hagan, J. (1988) 'Skills and Job Commitment in High Technology Industries in the US', *New Technology, Work and Employment* 3, 2: 112–124.

Hyman, R. (1989) *The Political Economy of Industrial Relations: Theory and Practice in a Cold Climate*, London: Macmillan.

IDA (1984) *Ireland: The New Hi-Tech Centre of Europe*, Dublin: Industrial Development Authority.

Irish Times (The) (1.12.1989) 'Emigration is Still a Reality for Ireland's Engineers'.

—— (9.11.1990) 'Number of Ph.D.s Going Abroad Rises'.

—— (11.1.1991) 'Electronics Now a Big Contributor to Irish Economy'.

—— (24.4.1991) 'The Reasons for the Lack of R&D'.

—— (25.6.1996) 'Another Record Year for Jobs'.

—— (19.7.1996) 'Irish Education and Financial Enticements Draw the Multinationals'.

Jacobsen, J. K. (1994) *Chasing Progress in the Irish Republic*, Cambridge: Cambridge University Press.

Jaques, E. (1967) *Equitable Payment*, London: Heinemann Educational Books (Revised edition).

Jenkins, R. (1982) 'Managers, Recruitment Procedures and Black Workers', *Working Papers on Ethnic Relations*, No.18, SSRC Research Unit on Ethnic Relations.

—— (1988) 'Discrimination and Equal Opportunity in Employment: Ethnicity and "Race" in the United Kingdom', in D. Gallie (ed.) *Employment in Britain*, Oxford: Basil Blackwell.

Jones, B., Scott, P., Bolton, B. and Bramley, A. (1993) 'Graduate Engineers and British Transnational Business: Elite Human Resource or Technical Labourers?', *Human Resource Management Journal* 4, 1: 34–49.

Jones, O. (1996) 'Human Resources, Scientists, and Internal Reputation', *Human Relations* 49, 3: 269–294.

Kanter, R. M. (1977) *Men and Women of the Corporation*, New York: Basic Books.

—— (1989) *When Giants Learn to Dance*, New York: Simon and Schuster.

Katz, R. (ed.) (1988) *Managing Professionals in Innovative Organizations*, Cambridge, Mass.: Ballinger.

Keating, W. and Keane, T. (1991) 'The Contraction and Expansion of Overseas Industry', in A. Foley and D. McAleese (eds) *Overseas Industry in Ireland*, Dublin: Gill and Macmillan.

Keenoy, T. and Anthony, P. D. (1992) 'HRM: Metaphor, Meaning and Morality', in P. Blyton and P. Turnbull (eds) *Reassessing Human Resource Management*, London: Sage.

Kelly, A. and Brannick, T. (1985) 'Industrial Relations Practices in Multinational Companies in Ireland', *Irish Business and Administrative Research* 7, 1: 98–111.

Kennedy, K., Giblin, T. and McHugh, D. (1988) *The Economic Development of Ireland in the Twentieth Century*, London: Routledge.

Kerr, S., Von Glinow, M.A. and Schriesheim, J. (1977) 'Issues in the Study of

"Professionals" in Organizations: The Case of Scientists and Engineers', *Organizational Behavior and Human Performance* 17: 329–345.

Kessler, I. and Purcell, J. (1992) 'Performance Related Pay: Objectives and Applications', *Human Resource Management Journal* 2, 3: 16–33.

Kornhauser, W. (1962) *Scientists in Industry*, Berkeley: University of California Press.

Kujawa, D. (1979) 'The Labour Relations of US Multinationals Abroad: Comparative and Prospective Views', *Labour and Society* 4, January: 3–25.

Lash, S. and Urry, J. (1987) *The End of Organized Capitalism*, Cambridge: Polity Press.

Legge, K. (1989) 'HRM : A Critical Analysis', in J. Storey, (ed.) *New Perspectives on Human Resource Management*, London: Routledge.

—— (1995) *Human Resource Management: Rhetorics and Realities*, Basingstoke: Macmillan.

Lincoln, J. R. and Kalleberg, A. L. (1990) *Culture Control and Commitment*, Cambridge: Cambridge University Press.

Lockwood, D. (1958) *The Blackcoated Worker*, Clarendon Press: Oxford.

—— (1989) 'Postscript', *The Blackcoated Worker*, 2nd edition, Clarendon Press: Oxford.

Lockwood, J., Teevan, P. and Walters, M. (1992) *Who's Managing the Managers? The Reward and Career Development of Middle Managers in a Flat Organization*, Corby: Institute of Management.

Long, F. (1980) 'Foreign Capital and Development Strategy in Irish Industrialization', *American Journal of Economics and Sociology* 39, 2: 137–150.

Lowe, J. and Oliver, N. (1991) 'The High-Commitment Workplace: Two Cases from a Hi-Tech Industry', *Work, Employment and Society* 5, 3: 437–450.

McAleese, D. and Counahan, M. (1979) '"Stickers" or "Snatchers"? Employment in Multinational Corporations During the Recession', *Oxford Bulletin of Economics and Statistics* 41, 4: 345–358.

McGovern, P. G. (1989) 'Union Recognition and Union Avoidance in the 1980s', in T. Murphy, B. Hillery and A. Kelly (eds) *Industrial Relations in Ireland: Contemporary Issues and Developments*, 2nd edition, Dublin: University College Dublin.

—— (1995) 'To Retain or Not to Retain? Technical Labour in Multinational firms', *Human Resource Management Journal* 5, 4: 7–23.

—— (1996a) 'Trust, Discretion and Responsibility: the Division of Technical Labour', *Work, Employment and Society* 10, 1: 85–103.

—— (1996b) 'Multinational Firms and Qualification Inflation in a Hi-Tech Labour Market', *New Technology, Work and Employment* 11, 1: 66–76.

McGovern, P. and Hope-Hailey, V. (1997) 'Inside Hewlett-Packard: Corporate Culture and Bureaucratic Control', in S. Sackmann (ed.) *Cultural Complexity in Organizations*, London: Sage.

McGovern, P., Hope-Hailey, V. and Stiles, P. (1998) 'The Managerial Career after Downsizing: Case Studies from the "Leading Edge"', *Work, Employment and Society* 12.

McGovern, P., Gratton, L., Hope-Hailey, V., Stiles, P. and Truss, C. (1997) 'Human Resource Management on the Line?', *Human Resource Management Journal* 7, 4: 12–29.

McGowan, K. (1991) 'The Electronics Industry', *The Irish Banking Review* Winter: 41–51.

Maguire, M. (1986) 'Recruitment as a Means of Control', in K. Purcell, S. Wood, A. Waton and S. Allen (eds) *The Changing Experience of Employment*, London: Macmillan.

Management (1985) 'Matching Invention with Markets', 32, 11: 6–9.

Marcson, S. (1969) *The Scientist in American Industry*, New York: Harper and Row.

Marshall, G., Newby, H., Rose, D. and Vogler, C. (1988) *Social Class in Modern Britain*, London: Hutchinson.

Meiksins, P. and Smith, C. (1993) 'Organizing Engineering Work: A Comparative Analysis', *Work and Occupations* 20, 2: 123–146.

—— (1996) *Engineering Labour: Technical Workers in Comparative Perspective*, London: Verso.

Miljus, R. and Smith, R. L. (1987) 'Key Human Resource Issues for Management in High Tech Firms', in A. Kleingartner and C. S. Anderson (eds) *Human Resource Management in High Technology Firms*, Lexington: Lexington Books.

Millward, N. (1994) *The New Industrial Relations*, London: Policy Studies Institute.

Mintzberg, H. (1973) *The Nature of Managerial Work*, New York: Harper and Row.

—— (1994) *The Rise and Fall of Strategic Planning*, New York: Free Press.

Mitchell, J. C. (1983) 'Case and Situation Analysis', *The Sociological Review* 31, 2: 187–211.

Morris, T., Lydka, H. and Fenton-O'Creevy, M. (1993) 'Can Commitment be Managed? A Longitudinal Analysis of Employee Commitment and Human Resource Policies', *Human Resource Management Journal* 3, 3: 21–42

Mowday, R. T., Steers, R. M. and Porter, L. M. (1979) 'The Measurement of Organizational Commitment', *Journal of Vocational Behaviour* 14, 2: 224–247.

Murray, P. and Wickham, J. (1987) 'The Last Boy's Club? Training, Education and Gender in Electronics Factories', Paper presented to the *Annual Conference of the Sociological Association of Ireland.*

Murray, S. *et al.* (1984) *Survey of Employee/Industrial Relations in Irish Private Sector Manufacturing*, Dublin: Industrial Development Authority.

NESC (1982) *A Review of Industrial Policy*, Report prepared by the Telesis

Consultancy Group, No. 64, Dublin: National Economic and Social Council.

—— (1985) *Manpower Policy in Ireland*, No. 82, Dublin: National Economic and Social Council.

Nicholson, N. and West, M. (1988) *Managerial Job Change: Men and Women in Transition*, Cambridge: Cambridge University Press.

O'Canainn, A. (1995) 'Herr Ingenieur or the Grease Monkey? How the Managerial Prospects of Engineers are Perceived', *Human Resource Management Journal* 5, 4: 74–92.

OECD (1964) *The Training of Technicians in Ireland*, Report prepared by J. P. O'Donnell, OECD: Paris.

Offe, C. (1976) *Industry and Inequality*, London: Edward Arnold.

O'Hearn, D. (1989) 'The Irish Case of Dependency: An Exception to the Exceptions?', *American Sociological Review* 54, August: 578–596.

O'Malley, E. (1985) 'The Problem of Late Industrialization and the Experience of the Republic of Ireland', *Cambridge Journal of Economics* 9: 141–154.

—— (1986) 'Foreign Owned Industry in Ireland: Performance and Prospects', *ESRI Medium Term Outlook 1986–1990*, Dublin: Economic and Social Research Institute.

—— (1987) *The Irish Engineering Industry: Strategic Analysis and Policy Recommendations*, Paper No. 134, Dublin: Economic and Social Research Institute.

Osterman, P. (1996) *Broken Ladders: Managerial Careers in the New Economy*, Oxford: Oxford University Press.

Oxenham, J. (1984) 'Education versus Qualifications', in J. Oxenham (ed.) *Employers, Jobs and Qualifications*, London: Allen and Unwin.

Perrons, D. C. (1981) 'The Role of Ireland in the New International Division of Labour: A Proposed Framework for Regional Analysis', *Regional Studies* 15, 2: 81–100.

—— (1986) 'Unequal Integration in Global Fordism: The Case of Ireland', in A. J. Scott and M. Storper (eds) *Production, Work, Territory: The Geographical Anatomy of Industrial Capitalism*, Boston: Allen and Unwin.

Perrow, C. (1970) *Organizational Analysis: A Sociological View*, London: Tavistock Publications.

Popper, K. (1959) *The Logic of Scientific Discovery*, London: Hutchinson.

Raelin, J. (1985) 'The Basis for the Professional's Resistance to Managerial Control', *Human Resource Management* 24, 2: 147–176.

Randle, K. and Rainnie, A. (1997) 'Managing Creativity, Maintaining Control: A Study in Pharmaceutical Research', *Human Resource Management Journal* 7, 2: 32–46.

Ritzer, G. (1993) *The McDonaldization of Society*, Thousand Oaks: Pine Forge Press.

Roberts, B. C., Loveridge, R., Gennard, J., Eason, J. V. and others (1972) *Reluctant Militants: A Study of Industrial Technicians*, London: Heinemann Educational Books.

Roche, W. K. (1991) 'Trust Dynamics and Organizational Integration: the Micro-Sociology of Alan Fox', *British Journal of Sociology* 42, 1: 95–113.

Roche, W. K. and Geary, J. (1996) 'Multinational Companies in Ireland: Adapting or Diverging from National Industrial Relations Practices and Traditions', *Irish Business and Administrative Research* 17: 14–31.

Roomkin, M. J. (ed.) (1989) *Managers as Employees: An International Comparison of the Changing Character of Managerial Employment*, New York: Oxford University Press.

Rose, M. (1994) 'Level of Strategy and Regimes of Control', in J. Rubery and F. Wilkinson (eds) *Employer Strategy and the Labour Market*, Oxford: Oxford University Press.

Rosenbaum, J. (1984) *Career Mobility in a Corporate Hierarchy*, New York: Academic Press.

Roy, D. (1952) 'Quota Restriction and Goldbricking in a Machine Shop', *American Journal of Sociology*, 57: 427–442.

Savage, M., Dickens, P. and Fielding, T. (1988) 'Some Social and Political Implications of the Contemporary Fragmentation of the "Service Class" in Britain', *International Journal of Urban and Regional Research* 12, 3: 455–475.

Savage, M., Barlow, J., Dickens, P. and Fielding, T. (1992) *Property, Bureaucracy and Culture: Middle-Class Formation in Contemporary Britain*, London: Routledge.

Schoenberger, E. (1989) Multinational Corporations and the New International Division of Labour: A Critical Appraisal', in S. Wood (ed.) *The Transformation of Work*, London: Unwin Hyman.

Sewell, G. and Wilkinson, B. (1992) '"Someone to Watch over Me": Surveillance, Discipline and the Just-in-Time Labour Process', *Sociology* 26, 2: 271–289.

Shepard, H. A. (1958) 'The Dual Hierarchy in Research', *Research Management* Autumn.

Sklair, L. (1988) 'Foreign Investment, Irish Development and the International Division of Labour', *Progress in Planning* 29, 3: 147–216.

—— (1998) 'Globalization and the Corporations: The Case of the California *Fortune* Global 500', *International Journal of Urban and Regional Research*.

Smith, C. (1987) *Technical Workers: Class, Labour and Trade Unionism*, London: Macmillan.

—— (1990) 'Technical Workers: A Class and Organizational Analysis', in S. R. Clegg (ed.) *Organization Theory and Class Analysis: New Approaches and New Issues*, New York: de Gruyter.

Smith, I. (1992) 'Reward Management and HRM', in P. Blyton and P. Turnbull (eds) *Reassessing Human Resource Management*, London: Sage.

Sofer, C. (1970) *Men in Mid-career: A Study of British Managers and Technical Specialists*, Cambridge: Cambridge University Press.

Steers, R. M. (1977) 'Antecedents and Outcomes of Organizational Commitment', *Administrative Science Quarterly* 22: 46–56.

Steiner, D. D. and Farr, J. L. (1986) 'Career Goals, Organizational Reward Systems and Technical Updating in Engineers', *Journal of Occupational Psychology* 59, 1: 13–24.

Storey, J. (ed.) (1989) *New Developments in Human Resource Management*, London: Routledge and Kegan Paul.

—— (1992) *Developments in the Management of Human Resources: An Analytical Review*, Oxford: Basil Blackwell.

Streeck, W. (1992) *Social Institutions and Economic Performance: Studies of Industrial Relations in Advanced Capitalist Economies*, London: Sage.

Thompson, P. and Ackroyd, S. (1995) 'All Quiet on the Workplace Front? A Critique of Recent Trends in British Industrial Sociology', *Sociology* 29, 4: 615–634.

Toner, W. P. (1987) 'Union or Non-union? Contemporary Employee Relations Strategies in the Republic of Ireland', unpublished Ph.D. thesis, London School of Economics.

Truss, C., Gratton, L., Hope-Hailey, V., McGovern, P. and Stiles, P. (1997) 'Soft and Hard Models of HRM: A Reappraisal', *Journal of Management Studies* 34, 1: 53–73.

Turbin, M. S. and Rosse, J. G. (1990) 'Staffing Issues in the High Technology Industry', in Luis R. Gomez-Mejia and Michael W. Lawless (eds) *Organization Issues in High Technology Management,* London: JAI Press.

Turner, T., D'Art, D. and Gunnigle, P. (1997) 'US Multinationals: Changing the Framework of Irish Industrial Relations?', *Industrial Relations Journal* 28, 2: 92–102.

Walton, R. E. (1985) 'From Control to Commitment in the Workplace', *Harvard Business Review* 63, 2: 77–84.

Watson, T. (1994) *In Search of Management*, London: Routledge.

Webb, J. (1992) 'The Mismanagement of Innovation', *Sociology* 26, 3: 471–492.

Weber, M. (1948) 'The Position of the Official', in H. H. Gerth and C. Wright Mills (eds) *From Max Weber*, London: Routledge and Kegan Paul.

Whalley, P. (1986a) *The Social Production of Technical Work: The Case of British Engineers*, London: Macmillan.

—— (1986b) 'Markets, Managers and Technical Autonomy', *Theory and Society* 15: 223–247.

Whalley, P. and Crawford, S. (1984) 'Locating Technical Workers in the Class Structure', *Politics and Society* 13, 3: 239–252.

Wheatley, M. (1992) *The Future of Middle Management*, Corby: Institute of Management.

Whelan, C. T. (1982) *Worker Priorities, Trust in Management and Prospects for Workers Participation*, Paper No. 111, Dublin: The Economic and Social Research Institute.

Whittington, R. (1990) 'The Changing Structures of R&D: From Centralization to Fragmentation', in R. Loveridge and M. Pitt (eds) *The Strategic Management of Technological Innovation*, London: John Wiley and Sons.

—— (1991a) 'The Fragmentation of Industrial R&D', in A. Pollert (ed.) *Farewell to Flexibility?*, Oxford: Basil Blackwell.

—— (1991b) 'Changing Control Strategies in Industrial R&D', *R&D Management* 21, 1: 43–53.

Whittington, R., McNulty, T. and Whipp, R. (1994) 'Market-Driven Change in Professional Services: Problems and Processes', *Journal of Management Studies*, 31, 3: 829–845.

Wickham, J. (1989) 'The Over Educated Engineer? The Work, Education and Careers of Irish Electronics Engineers', *Irish Business and Administrative Research* 10: 19–33.

—— (1992) 'Irish Engineers: Education for Emigration?', in G. Lee and C. Smith (eds) *Engineers and Management: International Comparisons*, London: Routledge.

Winstanley, D. (1991) 'Recruitment Strategies and Managerial Control of Technological Staff', in C. Smith, D. Knights and H. Willmott (eds) *White-Collar Work: The Non-Manual Labour Process*, London: Macmillan.

Wood, S. and Elliot, R. (1977) 'A Critical Evaluation of Fox's Radicalization of Industrial Relations Theory', *Sociology* 11: 101–125.

Wood, S. and Albanese, M. T. (1995) 'Can We Speak of a High Commitment Management on the Shop Floor?', *Journal of Management Studies* 32, 2: 215–247.

Wright, E. O. (1985) *Classes*, London: Verso.

Zussman, R. (1985) *Mechanics of the Middle Class: Work and Politics among American Engineers*, Berkeley: University of California Press.

INDEX